D1559646

ELECTRICAL

PROFESSIONAL REFERENCE

2014 CODE

Paul Rosenberg

Published by:

Australia • Brazil • Mexico • Singapore • United Kingdom • United States

www.DeWALT.com/GUIDES

OTHER TITLES AVAILABLE

This Book Belongs To:

Name:_____

Company: _____

Title: _____

Department: _____

Company Address: _____

_____

_____

Company Phone: _____

Home Phone: _____

CENGAGE
Learning·

DeWALT® Electrical Professional Reference: 2014 Code
Paul Rosenberg

SVP, GM Skills & Global Product Management:
Dawn Gerrain

Product Director:
Matthew Seeley

Product Team Manager:
James DeVoe

Product Manager:
Vanessa Myers

Senior Director, Development:
Marah Bellegarde

Senior Product Development Manager: Larry Main

Associate Content Developer:
Jenn Wheaton

Product Assistant:
Andrew Ouimet

Vice President, Marketing Services: Jennifer Ann Baker

Marketing Manager:
Scott Chrysler

Senior Production Director:
Wendy Troeger

Production Director:
Patty Stephan

Senior Content Project Manager: Stacey Lamodi

Senior Art Director:
Benjamin Gleeksman

For product information and technology assistance, contact us at **Cengage Learning Customer & Sales Support, 1-800-354-9706**

For permission to use material from this text or product, submit all requests online at **www.cengage.com/permissions**
Further permissions questions can be emailed to **permissionrequest@cengage.com**

Library of Congress Control Number: 2014951396

ISBN-13: 978-1-305-39506-0

Cengage Learning
20 Channel Center Street
Boston, MA 02210
USA

Cengage Learning is a leading provider of customized learning solutions with office locations around the globe, including Singapore, the United Kingdom, Australia, Mexico, Brazil, and Japan. Locate your local office at:
www.cengage.com/global

Cengage Learning products are represented in Canada by Nelson Education, Ltd.

Visit us at www.ConstructionEdge.cengage.com

For more learning solutions, to learn more about Cengage Learning, please visit **www.cengage.com**

Notice to the Reader

Printed in China

1 2 3 4 5 18 17 16 15

CONTENTS

CHAPTER 2 – Conduit, Cable, and Underground Installations 2-1

CHAPTER 4 – *Receptacles, Switches, Interior Wiring, and Lighting* 4-1

CHAPTER 9 – *Electrical and Job Safety* . . .9-1

CHAPTER 10 – *Materials and Tools* **10-1**

Preface

Many years ago I was asked to create a compendium of tables, charts, graphs, diagrams, and terminology relating to the field of electricity. Electrical Reference was my first endeavor to create a pocket guide for the electrical marketplace that would serve as an all encompassing publication in a new user-friendly format. I ended up utilizing a design and typeface that holds the most information per page while at the same time making a particular topic quick to reference.

Information covered in this manual is necessary for anyone in the field to have in one's possession at all times. Naturally, a topic may have been overlooked or not discussed in depth to suit all tradespeople. I will constantly monitor and update this book on a regular basis to not only include requested additional material, but to add new material from the ever growing amount of high technology as it develops.

Best wishes,
Paul Rosenberg

A Note To The Reader

Electrical Professional Reference is not a substitute for the National Electric Code®. NFPA 70®, *National Electrical Code,* and *NEC®* are registered trademarks of the National Fire Protection Association, Quincy, MA.

CHAPTER 1
Circuits, Formulas, and Voltage Drop Calculations

OHM'S LAW/POWER FORMULAS

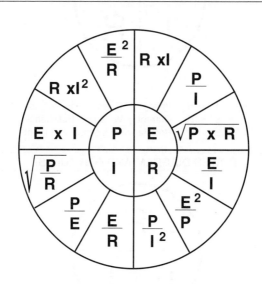

P = Power = Watts

R = Resistance = Ohms

I = Current = Amperes

E = Force = Volts

OHM'S LAW DIAGRAM AND FORMULAS

$E = I \times R$ Voltage = Current $\times$ Resistance
$I = E \div R$ Current = Voltage $\div$ Resistance
$R = E \div I$ Resistance = Voltage $\div$ Current

POWER DIAGRAM AND FORMULAS

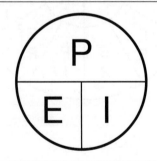

$P = I \times E$ Power = Current $\times$ Voltage
$E = P \div I$ Voltage = Power $\div$ Current
$I = P \div E$ Current = Power $\div$ Voltage

OHM'S LAW AND IMPEDANCE

E = Voltage (in volts)
I = Current (in amps)
Z = Impedance (in ohms)

$$E = I \times Z \qquad I = \frac{E}{Z} \qquad Z = \frac{E}{I}$$

Ohm's law and the power formula are limited to circuits in which electrical resistance is the only significant opposition to current flow including all DC circuits and AC circuits that do not contain a significant amount of inductance and/or capacitance. AC circuits that include inductance are any circuits that include a coil as the load, for example, motors, transformers, and solenoids. AC circuits that include capacitance are any circuits that include capacitors or other capacitive components.

Opposition to current flow in DC and AC circuits that do not contain a significant amount of inductance or capacitance is termed resistance (R). In circuits that do contain inductance (L) or capacitance (C), the opposition to the flow of current is termed reactance (X). In circuits that contain resistance (R) and reactance (X), the combined opposition to the flow of current is termed impedance (Z). Resistance, reactance, and impedance are all measured in ohms (Ω).

Ohm's law is used in circuits that contain impedance, however, and Z is substituted for R in the formula. Z represents the total resistive force (resistance and reactance) opposing current flow.

AC/DC POWER FORMULAS

To Find	Direct Current	Single-Phase	Two-Phase Four-Wire*	Three-Phase
		Alternating Current		
Amps when horsepower is known	$\dfrac{HP \times 746}{E \times E_{FF}}$	$\dfrac{HP \times 746}{E \times E_{FF} \times PF}$	$\dfrac{HP \times 746}{2 \times E \times E_{FF} \times PF}$	$\dfrac{HP \times 746}{1.73 \times E \times E_{FF} \times PF}$
Amps when kilowatts are known	$\dfrac{kW \times 1{,}000}{E}$	$\dfrac{kW \times 1{,}000}{E \times PF}$	$\dfrac{kW \times 1{,}000}{2 \times E \times PF}$	$\dfrac{kW \times 1{,}000}{1.73 \times E \times PF}$
Amps when kVA is known	—	$\dfrac{kVA \times 1{,}000}{E}$	$\dfrac{kVA \times 1{,}000}{2 \times E}$	$\dfrac{kVA \times 1{,}000}{1.73 \times E}$
Kilowatts	$\dfrac{I \times E}{1{,}000}$	$\dfrac{I \times E \times PF}{1{,}000}$	$\dfrac{I \times E \times 2 \times PF}{1{,}000}$	$\dfrac{I \times E \times 1.73 \times PF}{1{,}000}$
Kilovolt-Amps	—	$\dfrac{I \times E}{1{,}000}$	$\dfrac{I \times E \times 2}{1{,}000}$	$\dfrac{I \times E \times 1.73}{1{,}000}$
Horsepower	$\dfrac{I \times E \times E_{FF}}{746}$	$\dfrac{I \times E \times E_{FF} \times PF}{746}$	$\dfrac{I \times E \times 2 \times E_{FF} \times PF}{746}$	$\dfrac{I \times E \times 1.73 \times E_{FF} \times PF}{746}$

I = Amps; E = Volts; E_{FF} = Efficiency; PF = Power Factor; kW = Kilowatts; kVA = Kilovolt Amps; HP = Horsepower
*For three-phase, two-phase circuits the current in the common conductor is 1.41 times the current in either of the other two conductors.

Power Factor = cos (Phase Angle)
Power Factor = True Power / Apparent Power
Power Factor = Watts / Volts × Amps

Power Factor = Resistance (in ohms) / Impedance (in ohms)
Efficiency = Output / Input

1-4

THREE-PHASE AC CIRCUITS AND THE UTILIZATION OF POWER

Sine waves are actually an oscillograph trace taken at any point in a three-phase system. (Each voltage or current wave actually comes from a separate wire but are shown together for clarity.) There are 120° between consecutive voltages. At any instant the algebraic sum (measured up and down from centerline) of these three voltages is zero. When one voltage is zero, the other two are each 86.6% maximum and have opposite signs.

WYE CONNECTION

Consider the three windings as primary of transformer. Current in all windings equals line current, but volts across windings = 0.577 × line volts.

DELTA CONNECTION

Winding voltages equal line voltages, but currents split up so 0.577 I_{line} flows through windings.

FOUR-WIRE SYSTEM

Most popular secondary distribution setup. V_1 is usually 208 V, which feeds small power loads. Lighting loads at 120 V tap from any line to neutral.

MULTI-WIRE SYSTEM VOLTAGES

Wye-Connected

120/208 V, 3 φ, 4-wire wye

A to B = 208 V
B to C = 208 V
A to C = 208 V
A to N = 120 V
B to N = 120 V
C to N = 120 V

277/480 V, 3 φ, 4-wire wye

A to B = 480 V
B to C = 480 V
A to C = 480 V
A to N = 277 V
B to N = 277 V
C to N = 277 V

Delta-Connected

120/240 V, 3ϕ, 4-wire delta

A to B = 240 V
B to C = 240 V
A to C = 240 V
A to N = 120 V
B to N = 120 V
C to N = 208 V

480 V, 3ϕ, 3-wire delta

A to B = 480 V
B to C = 480 V
A to C = 480 V

FORMULAS FOR SINE WAVES

Frequency	Period	RMS Voltage
$F = \dfrac{1}{T}$ where F = Frequency (in hertz) T = Period (in seconds) 1 = Constant	$T = \dfrac{1}{F}$ where T = Period (in seconds) F = Frequency (in hertz) 1 = Constant	$V_{RMS} = V_{MAX} \times 0.707$ where V_{RMS} = RMS voltage V_{MAX} = Peak voltage 0.707 = Constant

Average Voltage	Peak-to-Peak Voltage
$V_{AVG} = V_{MAX} \times 0.637$ where V_{AVG} = Average voltage V_{MAX} = Peak voltage 0.637 = Constant	$V_{P-P} = V_{MAX} \times 2$ where V_{P-P} = Peak-to-peak voltage V_{MAX} = Peak voltage 2 = Constant

CALCULATING PEAK, AVERAGE, AND ROOT-MEAN-SQUARE (RMS) VOLTAGES

Effective (RMS) voltage = $0.707 \times$ Peak voltage
Effective (RMS) voltage = $1.11 \times$ Average voltage
Average voltage = $0.637 \times$ Peak voltage
Average voltage = $0.9 \times$ Effective (RMS) voltage
Peak voltage = $1.414 \times$ Effective (RMS) voltage
Peak voltage = $1.57 \times$ Average voltage
Peak-to-peak voltage = $2 \times$ Peak voltage
Peak-to-peak voltage = $2.828 \times$ Effective (RMS) voltage

KIRCHHOFF'S LAWS

First Law of Current

The sum of the currents arriving at any point in a circuit must equal the sum of the currents leaving that point.

Second Law of Voltage

The total voltage applied to any closed circuit path is always equal to the sum of the voltage drops in that path.

POWER FACTOR

An AC electrical system carries two types of power: (1) true power, measured in watts, that drives the load, and (2) reactive power, measured in VARs, that generates magnetism within inductive equipment. The vector sum of these two will represent the actual volt-amperes flowing in the circuit (see diagram below). Power factor is the cosine of the angle between true power and volt-amperes.

TYPES OF POWER

Heating element rated in true power (watts)

Transformer rated in apparent power (volt-amps)

Control transformer 460 V to 120 V

Capacitor in motor rated in reactive power (volt-amps reactive)

Reactive power is supplied to a reactive load (capacitor/coil) and is measured in volt-amps reactive (VAR). The capacitor on a motor uses reactive power to keep the capacitor charged. The capacitor uses no true power because it performs no actual work such as producing heat or motion.

In an AC circuit containing only resistance, the power in the circuit is true power. However, almost all AC circuits include capacitive reactance (capacitors) and/or inductive reactance (coils). Inductive reactance is the most common, because all motors, transformers, solenoids, and coils have inductive reactance.

Apparent power represents a load or circuit that includes both true power and reactive power and is expressed in volt-amps (VA), kilovolt-amps (kVA), or megavolt-amps (MVA). Apparent power is a measure of component or system capacity because apparent power considers circuit current regardless of how it is used. For this reason, transformers are sized in volt-amps rather than in watts.

TRUE POWER AND APPARENT POWER

True power is the actual power used in an electrical circuit and is expressed in watts (W). *Apparent power* is the product of voltage and current in a circuit calculated without considering the phase shift that may be present between total voltage and current in the circuit. Apparent power is measured in volt-amperes (VA). A phase shift exists in most AC circuits that contain devices causing capacitance or inductance.

True power equals apparent power in an electrical circuit containing only resistance. True power is less than apparent power in a circuit containing inductance or capacitance.

Capacitance is the property of a device that permits the storage of electrically separated charges when potential differences exist between the conductors. *Inductance* is the property of a circuit that causes it to oppose a change in current due to energy stored in a magnetic filed (i.e., coils).

To calculate true power, apply the formula:

$$P_T = I^2 \times R$$

where

P_T = true power (in watts)
I = total circuit current (in amps)
R = total resistive component of the circuit (in ohms)

To calculate apparent power, apply the formula:

$$P_A = E \times I$$

where

P_A = apparent power (in volt-amps)
E = measured voltage (in volts)
I = measured current (in amps)

POWER FACTOR FORMULA

Power factor is the ratio of true power used in an AC circuit to apparent power delivered to the circuit.

$$PF = \frac{P_T}{P_A}$$

where

PF = power factor (percentage)
P_T = true power (in watts)
P_A = apparent power (in volt-amps)

POWER FACTOR IMPROVEMENT

Capacitor Multipliers for Kilowatt Load

To give capacitor kvar required to improve power factor from original to desired value. For example, assume the total plant load is 100kw at 60 percent power factor. Capacitor kvar rating necessary to improve power factor to 80 percent is found by multiplying kw (100) by multiplier in table (0.583), which gives kvar (58.3). Nearest standard rating (60 kvar) should be recommended.

Original Power Factor Percentage	Desired Power Factor Percentage				
	100%	95%	90%	85%	80%
60	1.333	1.004	0.849	0.713	0.583
62	1.266	0.937	0.782	0.646	0.516
64	1.201	0.872	0.717	0.581	0.451
66	1.138	0.809	0.654	0.518	0.388
68	1.078	0.749	0.594	0.458	0.338
70	1.020	0.691	0.536	0.400	0.270
72	0.964	0.635	0.480	0.344	0.214
74	0.909	0.580	0.425	0.289	0.159
76	0.855	0.526	0.371	0.235	0.105
77	0.829	0.500	0.345	0.209	0.079
78	0.802	0.473	0.318	0.182	0.052
79	0.776	0.447	0.292	0.156	0.026
80	0.750	0.421	0.266	0.130	
81	0.724	0.395	0.240	0.104	
82	0.698	0.369	0.214	0.078	
83	0.672	0.343	0.188	0.052	
84	0.646	0.317	0.162	0.026	
85	0.620	0.291	0.136		
86	0.593	0.264	0.109		
87	0.567	0.238	0.083		
88	0.540	0.211	0.056		
89	0.512	0.183	0.028		
90	0.484	0.155			
91	0.456	0.127			
92	0.426	0.097			
93	0.395	0.066			
94	0.363	0.034			
95	0.329				
96	0.292				
97	0.251				
99	0.143				

CAPACITOR CORRECTION FOR THREE-PHASE MOTORS

HP	600 RPM Capacitor Rating KVAR	600 RPM % of Current Reduced	900 RPM Capacitor Rating KVAR	900 RPM % of Current Reduced	1,200 RPM Capacitor Rating KVAR	1,200 RPM % of Current Reduced	1,800 RPM Capacitor Rating KVAR	1,800 RPM % of Current Reduced	3,600 RPM Capacitor Rating KVAR	3,600 RPM % of Current Reduced
10	7.5	31	5.0	21	3.5	14	3.0	11	3.0	10
15	9.5	27	6.5	18	5.0	13	4.0	10	4.0	9
20	12.0	25	7.5	16	6.5	12	5.0	10	5.0	9
25	14.0	23	9.0	15	7.5	11	6.0	10	6.0	9
30	16.0	22	10.0	14	9.0	11	7.0	9	7.0	8
40	20.0	20	12.0	13	11.0	10	9.0	9	9.0	8
50	24.0	19	15.0	12	13.0	10	11.0	9	12.0	8
60	27.0	19	18.0	11	15.0	10	14.0	8	14.0	8
75	32.5	18	21.0	10	18.0	10	16.0	8	17.0	8
100	40.0	17	27.0	10	25.0	9	21.0	8	22.0	8
125	47.5	16	32.5	10	30.0	9	26.0	8	27.0	8
150	52.5	15	37.5	10	35.0	9	30.0	8	32.5	8
200	65.0	14	47.5	10	42.5	9	37.5	8	40.0	8

Motors shown are open-type induction at 60 hertz.
For lower rated capacitors, reduce proportionately to line current.

DELTA AND WYE RESISTOR CIRCUITS

In the delta network the resistance between terminals may be determined by combining the formulas for series and parallel resistances as follows:

$$a \text{ to } b = \frac{R_z \times (R_x + R_y)}{R_z + (R_x + R_y)}$$

$$a \text{ to } c = \frac{R_y \times (R_x + R_z)}{R_y + (R_x + R_z)}$$

$$b \text{ to } c = \frac{R_x \times (R_z + R_y)}{R_x + (R_z + R_y)}$$

Delta

To convert from delta to wye, use:

$$R_a = \frac{R_y \times R_z}{R_x + R_y + R_z}$$

$$R_b = \frac{R_x \times R_z}{R_x + R_y + R_z}$$

$$R_c = \frac{R_x \times R_y}{R_x + R_y + R_z}$$

Wye

To convert from wye to delta, use:

$$R_x = \frac{(R_a \times R_b) + (R_b \times R_c) + (R_c \times R_a)}{R_a}$$

$$R_y = \frac{(R_a \times R_b) + (R_b \times R_c) + (R_c \times R_a)}{R_b}$$

$$R_z = \frac{(R_a \times R_b) + (R_b \times R_c) + (R_c \times R_a)}{R_c}$$

SUMMARY OF SERIES, PARALLEL, AND COMBINATION CIRCUITS

To Find	Series Circuits	Parallel Circuits	Series/Parallel
Resistance (R)	$R_T = R_1 + R_2 + R_3$ Sum of individual resistances	$\frac{1}{R_T} = \frac{1}{R_1} + \frac{1}{R_2} + \frac{1}{R_3}$	Total resistance equals resistance of parallel portion and sum of series resistors
Current (I)	$I_T = I_1 = I_2 = I_3$ The same throughout entire circuit	$I_T = I_1 + I_2 + I_3$ Sum of individual currents	Series rules apply to series portion of circuit and parallel rules apply to parallel part of circuit
Voltage (E)	$E_T = E_1 + E_2 + E_3$ Sum of individual voltages	$E_T = E_1 = E_2 = E_3$ Total voltage and branch voltage are the same	Total voltage is sum of voltage drops across each series resistor and each of the branches of parallel portion
Power (P)	$P_T = P_1 + P_2 + P_3$ Sum of individual wattages	$P_T = P_1 + P_2 + P_3$ Sum of individual wattages	$P_T = P_1 + P_2 + P_3$ Sum of Individual wattages

CIRCUIT CHARACTERISTICS

Resistances in a Series DC Circuit

10 V

4 Ω   6 Ω

$R_T = 10\ \Omega$

Parallel Circuit, Showing Voltage Drops

20 V

20 V
R_1 2 Ω

20 V
R_2 3 Ω

20 V
R_3 4 Ω

Voltages in a Series-Parallel Circuit

Parallel Circuit, Showing Current Values

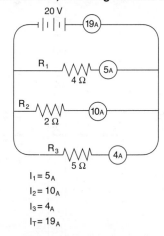

$I_1 = 5_A$

$I_2 = 10_A$

$I_3 = 4_A$

$I_T = 19_A$

CAPACITANCE

Capacitors in Series

$$C_T = \frac{C_1 \times C_2}{C_1 + C_2} \quad \text{(Two only)}$$

$$\frac{1}{C_T} = \frac{1}{C_1} + \frac{1}{C_2} + \frac{1}{C_3} + \cdots \text{ (Multiple)}$$

Capacitors in Parallel

$$C_T = C_1 + C_2 + C_3 + \cdots$$

CAPACITORS CONNECTED IN SERIES/PARALLEL

1. Calculate the capacitance of the parallel branch.

$$C_{PT} = C_1 + C_2 + C_3 + \cdots$$

2. Calculate the capacitance of the series combination.

$$C_T = \frac{C_{PT} \times C_S}{C_{PT} + C_S}$$

CAPACITIVE REACTANCE

Capacitive reactance (X_C) is the opposition to current flow by a capacitor when connected to an AC power supply and is expressed in ohms. To calculate capacitive reactance, apply the formula:

$$X_C = \frac{1}{2\pi fc}$$

where

X_C = capacitive reactance (in ohms)

2π = 6.28 (indicates circular motion that produces the AC sine wave)

f = applied frequency (in hertz)

c = capacitance (in farads)

Capacitive Reactances

In Series   In Parallel

$X_{CT} = X_{C1} + X_{C2} + X_{C3} \cdots$ $X_{CT} = \dfrac{X_{C_1} \times X_{C_2}}{X_{C_1} + X_{C_2}}$ (only two)

(multiple)

$$\frac{1}{X_{C_T}} = \frac{1}{X_{C_1}} + \frac{1}{X_{C_2}} + \frac{1}{X_{C_3}} + \cdots$$

To calculate capacitive reactance when voltage across the capacitor (E_C) and current through the capacitor (I_C) are known, apply the formula:

$$X_C = \frac{E_C}{I_C}$$

where

X_C = capacitive reactance (in ohms)

E_C = voltage across capacitor (in volts)

I_C = current through capacitor (in amps)

INDUCTANCE

Inductors in Series

$$L_T = L_1 + L_2 + L_3 + \cdots$$

Inductors in Parallel

$$L_T = \frac{L_1 \times L_2}{L_1 + L_2} \quad \text{(only two)}$$

$$\frac{1}{L_T} = \frac{1}{L_1} + \frac{1}{L_2} + \frac{1}{L_3} + \cdots \quad \text{(multiple)}$$

INDUCTORS CONNECTED IN SERIES/PARALLEL

(only two)

$$L_T = \left(\frac{L_1 \times L_2}{L_1 + L_2} \right) + L_1 + L_2 + L_3 + \cdots$$

parallel   series

(multiple)

$$L_T = \left(\frac{1}{\dfrac{1}{L_1} + \dfrac{1}{L_2} + \dfrac{1}{L_3} + \cdots} \right) + L_1 + L_2 + L_3 + \cdots$$

parallel   series

INDUCTIVE REACTANCE

Inductive reactance (X_L) is an inductor's opposition to alternating current measured in ohms.

To calculate inductive reactance for AC circuits, apply the formula:

$$X_L = 2\pi fL$$

where

X_L = inductive reactance (in ohms)
$2\pi = 6.28$ (indicates circular motion that produces the AC sine wave)
f = applied frequency (in hertz)
L = inductance (in henrys)

Inductive Reactances

In Series

$$X_{L_T} = X_{L_1} + X_{L_2} + \cdots$$

In Parallel

$$X_{LT} = \frac{X_{L_1} \times X_{L_2}}{X_{L_1} + X_{L_2}} \qquad \text{(only two)}$$

$$\frac{1}{X_{L_T}} = \frac{1}{X_{L_1}} + \frac{1}{X_{L_2}} + \frac{1}{X_{L_3}} \cdots \text{(multiple)}$$

To calculate inductive reactance when voltage across a coil (E_L) and current through a coil (I_L) are known.

$$X_L = \frac{E_L}{I_L}$$

where

X_L = inductive reactance (in ohms)
E_L = voltage across coil (in volts)
I_L = current through coil (in amps)

VOLTAGE DROP USING OHM'S LAW

The definition of Ohm's Law is: Voltage equals current (amperes) times resistance (ohms).

A working model using the following criteria will calculate the voltage drop in a power line. The line is 300 feet long (150' and back) using 12-guage copper wire powering an 800-watt light fixture from a 120-volt supply.

Resistance of #12 wire is 0.00162 ohms/foot

Current is 800 watts/120 volts = 6.67 amps

Resistance is 300 feet $\times$ 0.00162 ohms/foot = 0.486 ohms

Voltage Drop = 6.67 amps $\times$ 0.486 ohms = 3.24 volts

The Voltage Drop in percentage = 3.24 volts/120 volts = 2.7%

USING DISTANCE TO LOAD

For a 1ϕ, 2-wire, line-to-line or line-to-neutral	Use $E = I \times R \times 2$
For a 1ϕ, 3-wire, line-to-neutral with a balanced neutral	Use $E = I \times R$
For a 3ϕ, 3-wire, line-to-line	Use $E = I \times R \times 1.73$
For a 3ϕ, 4-wire, line-to-neutral with a balanced neutral	Use $E = I \times R$

E = drop in volts (to calculate in %, divide E by the circuit voltage)
I = amperage of the load
R = resistance of the conductors based on length in one direction

CALCULATING BRANCH CIRCUIT VOLTAGE DROP IN PERCENT

$$\% V_D = \frac{V_{NL} - V_{FL}}{V_{NL}} \times 100$$

where
$\% V_D$ = percent voltage drop
V_{NL} = no-load voltage (volts)
V_{FL} = full-load voltage (volts)

VOLTAGE DROP FORMULAS

The NEC® recommends a maximum 3% voltage drop for either the branch circuit or the feeder.

Single-phase:

$$VD = \frac{2 \times R \times I \times L}{CM}$$

Three-phase:

$$VD = \frac{1.732 \times R \times I \times L}{CM}$$

VD = volts (voltage drop of the circuit)

R = 12.9 ohms/copper or 21.2 ohms/aluminum (resistance constants for a 1,000 circular mils conductor that is 1,000 ft long, at an operating temperature of 75°C)

I = amps (load at 100 %)

L = feet (length of circuit from load to power supply)

CM = circular-mils (conductor wire size)

2 = single-phase constant

1.732 = three-phase constant

CONDUCTOR LENGTH/VOLTAGE DROP

Voltage drop can be reduced by limiting the length of the conductors.

Single-phase:

$$L = \frac{CM \times VD}{2 \times R \times I}$$

Three-phase:

$$L = \frac{CM \times VD}{1.732 \times R \times I}$$

CONDUCTOR SIZE/VOLTAGE DROP

Increase the size of the conductor to decrease the voltage drop of circuit (reduce its resistance).

Single-phase:

$$CM = \frac{2 \times R \times I \times L}{VD}$$

Three-phase:

$$CM = \frac{1.732 \times R \times I \times L}{VD}$$

WIRE LENGTH VS. WIRE SIZE (MAX VOLTAGE DROP)

Max Wire Ft @ 240 V, Single-Phase, 2% Max Voltage Drop

3/0	2/0	1/0	#2	#4	Watts	Amps
180	—	—	—	—	48,000	200
240	190	185	—	—	36,000	150
360	280	230	—	—	24,000	100
440	365	290	180	—	19,200	80
520	415	330	205	130	16,800	70
600	485	385	240	150	14,400	60
720	580	460	290	180	12,000	50
880	725	575	360	230	5,600	40
1,200	970	770	485	300	7,200	30
1,440	1,100	920	580	365	6,000	25

#6	#8	#10	#12	#14	Watts	Amps
105	—	—	—	—	12,000	50
130	90	—	—	—	5,600	40
175	120	75	—	—	7,200	30
210	144	90	—	—	6,000	25
265	180	110	70	—	4,800	20
350	240	150	95	60	3,600	15
525	360	225	140	90	2,400	10
1,020	720	455	285	180	1,200	5

Max Wire Ft @ 120 V, Single-Phase, 2% Max Voltage Drop

3/0	2/0	1/0	#2	#4	Watts	Amps
230	180	144	90	—	9,600	80
260	205	165	105	65	8,400	70
305	240	190	120	76	7,200	60
360	290	230	145	90	6,000	50
440	360	290	175	115	4,800	40
600	490	385	240	150	3,600	30
720	580	460	290	180	3,000	25
900	725	575	365	230	2,400	20
1,200	965	770	485	305	1,800	15

#6	#8	#10	#12	#14	Watts	Amps
57	—	—	—	—	6,000	50
72	45	—	—	—	4,800	40
95	60	38	—	—	3,600	30
115	72	45	—	—	3,000	25
140	90	57	36	—	2,400	20
190	120	75	47	30	1,800	15
285	180	115	70	45	1,200	10
575	360	225	140	90	600	5

VOLTAGE DROP AMPERE-FEET

Copper Conductors, 70°C Copper Temp.,
600 V Class Single Conductor Cables in Steel Conduit

Maximum circuit ampere-feet without exceeding specified percentage voltage drop, various circuit voltages, and power factors.

Conductor Size	DC Circuits 1% Drop on 120 V	60 Cycle AC Circuits				
		1% Drop, 1.00 PF		3% Drop, 0.85 PF		
		120 V 1-Phase	208 V 3-Phase	115 V 1-Phase	208 V 3-Phase	220 V 3-Phase
14	191	191	382	623	1,300	1,380
12	305	305	612	998	2,080	2,200
10	484	484	968	1,580	3,280	3,470
8	770	770	1,540	2,450	5,110	5,410
6	1,200	1,190	2,380	3,800	7,850	8,310
4	1,900	1,890	3,780	5,750	12,000	12,700
2	3,030	2,970	5,950	8,620	18,000	19,100
1	3,820	3,710	7,430	10,400	21,800	23,100
1/0	4,820	4,560	9,120	12,500	26,000	27,500
2/0	6,060	5,610	11,200	14,800	30,800	32,700
3/0	7,650	6,940	13,900	16,900	35,200	37,100
4/0	9,760	8,520	17,100	20,200	42,100	44,600
250	11,500	9,930	19,800	22,300	46,500	49,300
300	13,800	11,500	23,100	24,800	52,000	55,000
350	16,100	13,200	26,300	27,000	56,200	59,500
400	18,400	14,000	28,100	28,700	60,000	63,500
500	22,700	16,400	32,800	31,800	66,400	70,300
750	34,600	20,400	40,800	36,700	76,600	81,200

Note: Length to be used is the distance from point of supply to load.

VOLTAGE DROP TABLE

Conductor Size	Voltage Drop per 1,000 Ampere-Feet							
		AC System						
		Load Power Factor (%)						
	DC	100	95	90	85	80	75	70

For DC circuit or single phase, 60-cycle, 2-wire system or 3-wire system with balanced load. Copper conductors, 70°C copper temperature. 600 V class single-conductor cables in steel conduit.

Conductor Size	DC	100	95	90	85	80	75	70
14	6.29	6.29	6.06	5.78	5.54	5.26	4.97	4.74
12	3.93	3.93	3.81	3.64	3.46	3.29	3.13	2.95
10	2.48	2.48	2.44	2.31	2.19	2.08	1.96	1.85
8	1.56	1.56	1.51	1.47	1.41	1.34	1.27	1.20
6	0.999	1.011	0.987	0.953	0.918	0.872	0.826	0.774
4	0.631	0.635	0.641	0.624	0.600	0.578	0.554	0.528
2	0.396	0.404	0.418	0.413	0.400	0.386	0.372	0.358
1	0.314	0.323	0.356	0.337	0.330	0.322	0.311	0.300
1/0	0.249	0.263	0.280	0.282	0.277	0.269	0.263	0.255
2/0	0.198	0.214	0.233	0.236	0.233	0.230	0.226	0.222
3/0	0.157	0.173	0.196	0.206	0.204	0.200	0.194	0.188
4/0	0.123	0.141	0.163	0.170	0.171	0.170	0.169	0.166
250	0.1041	0.121	0.146	0.152	0.155	0.155	0.155	0.154
300	0.0870	0.1040	0.128	0.135	0.139	0.140	0.141	0.141
350	0.0746	0.0912	0.117	0.125	0.128	0.131	0.131	0.131
400	0.0652	0.0855	0.1086	0.117	0.120	0.122	0.124	0.125
500	0.0528	0.0733	0.0959	0.1040	0.1086	0.111	0.113	0.114
750	0.0347	0.0589	0.0808	0.0884	0.0940	0.0976	0.0999	0.1020

For 3-phase, 60-cycle, 3-wire or 4-wire balanced system. Copper conductors, 70°C copper temperature. 600 V class single-conductor cables in steel conduit.

Conductor Size	DC	100	95	90	85	80	75	70
14		5.45	5.25	5.00	4.80	4.55	4.30	4.10
12		3.40	3.30	3.15	3.00	2.85	2.70	2.55
10		2.15	2.10	2.00	1.90	1.80	1.70	1.60
8		1.35	1.31	1.27	1.22	1.16	1.10	1.04
6		0.875	0.855	0.825	0.795	0.755	0.715	0.670
4		0.550	0.555	0.540	0.520	0.500	0.480	0.457
2		0.350	0.362	0.358	0.346	0.334	0.322	0.310
1		0.280	0.308	0.292	0.286	0.279	0.269	0.260
1/0		0.228	0.242	0.244	0.240	0.233	0.228	0.221
2/0		0.185	0.202	0.204	0.202	0.199	0.196	0.192
3/0		0.150	0.170	0.178	0.177	0.173	0.168	0.163
4/0		0.122	0.141	0.147	0.148	0.147	0.146	0.144
250		0.105	0.126	0.132	0.134	0.134	0.134	0.133
300		0.0900	0.111	0.117	0.120	0.121	0.122	0.122
350		0.0790	0.101	0.108	0.111	0.113	0.114	0.114
400		0.0740	0.0940	0.101	0.104	0.106	0.107	0.108
500		0.0635	0.0830	0.0900	0.0940	0.0964	0.0974	0.0988
750		0.0510	0.0700	0.0765	0.0814	0.0845	0.0865	0.0883

Note: Length to be used is the distance from point of supply to load, not amount of wire in circuit.

CHAPTER 2
Conduit, Cable, and Underground Installations

Double bend

Single bend

Change in direction

Offset

Kick

Bend

90° 90°

90°

90°

ON

OFF

Total of 360° of bends is permitted.

90° 90°

90°

90°

90°

ON

OFF

Over 360° of bends is not permitted.

BENDING EMT STUB-UPS

EMT Size (inches)	Take-up Amount (inches)
½	5
¾	6
1	8
1¼	11

Step One:
Determine the height of the stub-up required, and mark on the EMT.

Step Two:
Subtract the take-up amount from the stub height, and mark on the EMT that distance from the end.

Step Three:
Align the arrow on bender with the last mark made on the EMT, and bend to the 90° mark on the bender.

Making a 10" Stub from 1/2" EMT

BACK-TO-BACK BENDING

A back-to-back bend results in a U-shaped length of conduit, usually for runs along the floor or ceiling which then turn up or down a wall.

Step One:
After the first 90° bend is made, determine the back-to-back length and mark it on the EMT.

Step Two:
Align this back-to-back mark with the star on the bender, and bend to 90°.

40"

46"

2nd Mark or Arrow

40"

40"

2-BEND OFFSET

2-Bend Offset

Marks line up at arrow of bender

Angle of Bends	Inches of Run per Inch of Offset	Loss of Conduit Length per Inch of Offset
10°	5.76	1/16
22.5°	2.6	3/16
30°	2.0	1/4
45°	1.414	3/8
60°	1.15	1/2

2-BEND OFFSET *(cont.)*

An offset bend is used to change the level, or plane, of the conduit, usually because of an obstruction in the original conduit path.

Step One:
Determine the offset depth.

Step Two:
Multiply the offset depth by the multiplier for the degree of bend used to determine the distance between bends. 30° is the usual choice, because of the convenient 2.0 multiplier.

For example, if the offset depth required is 8", and you intend to use 30° bends, the distance between bends is 8" $\times$ 2 = 16".

Step Three:
Mark at the appropriate points, align the arrow on the bender with the first mark, and bend to desired degree by aligning the EMT with chosen degree line on the bender.

Step Four:
Slide down the EMT, align the arrow with the second mark, and bend to the same degree line. Be aware of the position of the bender head.

Degree line

Degree line

3-BEND SADDLE

45° Bend — Center of Bend
Center of Bend — Center of Bend
22.5° Bend

The 3-bend saddle is used when encountering an obstacle.

Step One:
Measure the height of the obstruction.
Mark the center point on the EMT.

Step Two:
Multiply the height of the obstruction by 2.5 and mark this distance on each side of the center mark, shown as D_2 and D_3.

Mark Center First

$\leftarrow$ D_2 $\rightarrow$ $\leftarrow$ D_3 $\rightarrow$
2nd Mark   3rd Mark

Step Three:
Place the center mark on the saddle mark/notch. Bend to 45°.

Step Four:
Bend the second mark to 22½° angle at arrow.

Step Five:
Bend the third mark to 22½° angle at arrow. Be sure to note the position of the EMT on all bends.

2nd Mark   3rd Mark

Center mark on
saddle mark / notch

CHICAGO-TYPE CONDUIT BENDER — 90° BENDING

A to C = Stub-up
C to D = Tail
C = Back of stub up
C = Bottom of conduit

There are normally two sizes of bending shoes. The *small* shoe takes ½", ¾", and 1" conduit. The *large* shoe takes 1¼" and 1½" conduit.

To determine the "take-up" and "shrink" of each size conduit for any bender to make 90° bends, use a straight piece of conduit and:

1) Measure length of conduit, A to D.
2) Place conduit in bender and mark at the edge of the shoe, B.
3) Level conduit. Bend ninety, and count number of pumps. Be sure to note each size of conduit you have used.
4) After bending you can find:
   A. Distance between B and C is the take-up.
   B. Original measurement of the piece of conduit subtracted from distance A to C plus distance C to D is the shrink.

Notes: Time can be saved if conduit can be cut, reamed, and threaded before you begin bending. Also, this method can be used on hydraulic benders as well.

CHICAGO-TYPE CONDUIT BENDER— OFFSET BENDING

Use a Straight Edge to Ensure Accuracy

Point A

6"

3" 4"

A

B

C 4"

D

In order to bend a 6" offset

1. Make a mark 3" from the conduit end. Place the conduit in the bender with the mark at the outside edge of the jaw.
2. Make three full pumps, making sure handle goes all the way down.
3. Remove the conduit from the bender and place it alongside a straight-edge.
4. Measure 6" from the straight-edge to the center of the conduit. Mark it point D. Use a square to ensure accuracy.
5. Mark the center of the conduit from both directions through bend as shown by the broken line. Where the lines intersect is point B.
6. Measure from A to B to determine the distance from D to C. Mark C and place the conduit in the bender with the mark at the outside edge of the jaw, and with the kick pointing down. Use a level to ensure accuracy.
7. Make three full pumps, making sure the handle goes all the way down.

MULTI-SHOT 90° CONDUIT BENDING

Always measure, thread, cut, and ream conduit before bending. The information contained in the following pages will take you through an example of multi-shot 90° conduit bending in a step-by-step process. The parameters of the example are as follows:

Size of the conduit = 2"
Space between conduits (center to center) = 6"
Height of the stub-up = 36"
Length of the tail = 48"

1) Determine the Radius (R):

   Conduit #1 will use the minimum radius, which is 8 times the size of the conduit.

   Thus: Radius of Conduit #1 = $8 \times 2" + 1.25" = 17.25"$
   Radius of Conduit #2 = $R_1 + 6" = 23.25"$
   Radius of Conduit #3 = $R_2 + 6" = 29.25"$

2) Determine the Developed Length (DL):

   Radius $\times$ 1.57 = DL

   Thus: DL of Conduit #1 = $R_1 \times 1.57 = 17.25" \times 1.57 = 27"$
   DL of Conduit #2 = $R_2 \times 1.57 = 23.25" \times 1.57 = 36.5"$
   DL of Conduit #3 = $R_3 \times 1.57 = 29.25" \times 1.57 = 46"$

3) Determine the Length of the Nipple:

   Thus:

   Nipple Length of Conduit #1 = $(L_1 + H_1 + DL_1) - (2 \times R_1)$
   $= (48" + 36" + 27") - (34.5") = 76.5"$
   Nipple Length of Conduit #2 = $(L_2 + H_2 + DL_2) - (2 \times R_2)$
   $= (54" + 42" + 36.5") - (46.5") = 86"$
   Nipple Length of Conduit #3 = $(L_3 + H_3 + DL_3) - (2 \times R_3)$
   $= (60" + 48" + 46") - (58.5") = 95.5"$

Note: For 90° bends, the shrink = $(2 \times R) - DL$
   For offset bends, the shrink = Hypotenuse − Side Adjacent

MULTI-SHOT 90° CONDUIT BENDING *(cont.)*

For the layout and the bending, follow the procedures below:

1) Locate point B by measuring from point A, the Height of the stub-up (H_1) minus the radius (R_1). This distance would be 18.75" on all three conduits.
2) Locate point C by measuring from point D, the Length (L_1) minus the radius (R_1). This distance would be 30.75" on all three conduits.
3) Divide the Developed Length (DL), which is point B to point C, into equal spaces not greater than 1.75" in order to prevent any possible wrinkling of the conduit. On Conduit #1, 17 spaces of 1.5882" each would result in 18 shots of 5° each. Note that there is always one less space than there are shots. When you are determining the quantity of shots, select a number that will divide evenly into 90.
4) An alternative method using a table corner and a ruler can be utilized as shown below:

A to B = Conduit #1
Developed Length or
DL (B to C) = 27"

A to C = 17 1" spaces
A to B = 17 1.5882" spaces
C = table corner

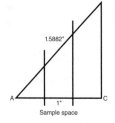

Sample space

Measure from Point C (table corner) 17 inches along table edge to Point A and mark. Place end of rule at Point A. Point B will be located where the 27" mark meets table edge B–C. Mark on board, then transfer to conduit.

MULTI-SHOT 90° CONDUIT BENDING *(cont.)*

$(L + H + DL) - (2 \times R) = \text{Nipple}$
$R \times 1.57 = DL$
$H - R = B$
$L - R = C$

TO LOCATE POINT B

$B = H_1 - R_1$
$= 36" - 17.25"$
$= 18.75"$

$B = H_2 - R_2$
$= 42" - 23.25"$
$= 18.75"$

$B = H_3 - R_3$
$= 48" - 29.25"$
$= 18.75"$

TO LOCATE POINT C

$C = L_1 - R_1$
$= 48" - 17.25"$
$= 30.75"$

$C = L_2 - R_2$
$= 54" - 23.25"$
$= 30.75"$

$C = L_3 - R_3$
$= 60" - 29.25"$
$= 30.75"$

Note that points B and C are equidistant from the end on all three conduits.

BOX OFFSETS					
Bend Angle	**Center-to-Center Distance**				
	¼"	⅜"	½"	⅝"	¾"
5°	2⅞"	4⁵⁄₁₆"	5¾"	7³⁄₁₆"	8⅝"
10°	1⁷⁄₁₆"	2³⁄₁₆"	2⅞"	3⁹⁄₁₆"	4⁵⁄₁₆"

OFFSET TRAVEL					
Bend Angle	**Travel in Inches**				
	½"	¾"	1"	1¼"	1½"
5°	¼"+	⁷⁄₁₆"−	⁹⁄₁₆"	¾"	1"
10°	½"−	1¹⁄₁₆"	1³⁄₁₆"	1⅜"	1¹¹⁄₁₆"
15°	¹¹⁄₁₆"	1¼"+	1⁹⁄₁₆"	2³⁄₁₆"	2⅝"
20°	¹⁵⁄₁₆"	1⅝"	2⅛"	2¹³⁄₁₆"	3⁷⁄₁₆"
22.5°	1¹⁄₁₆"	1⅞"	2⅜"	3³⁄₁₆"	3¹¹⁄₁₆"
30°	1⁷⁄₁₆"	2⁷⁄₁₆"+	3¹⁄₁₆"	4⅛"	4¾"
40°	1⅞"	3⁵⁄₁₆"	4⅛"	5⅜"	5½"
45°	2⅛"	3⅝"	4⅜"	6"	6¼"
60°	2¹³⁄₁₆"+	4⅞"	6¼"	8⅛"	9⅜"
90°	4³⁄₁₆"	7⁷⁄₁₆"	9⅜"	12¹⁄₁₆"	13¹³⁄₁₆"

KICK ADJUSTMENTS					
Bend Angle	**Adjustment in Inches**				
	½"	¾"	1"	1¼"	1½"
20°	1¹⁹⁄₃₂"	2¹³⁄₃₂"	2¹¹⁄₁₆"	2¹³⁄₁₆"	3⅜"
22.5°	1⅝"	2½"	2¹⁹⁄₃₂"	3"	3⅝"
30°	1²⁷⁄₃₂"	2¹¹⁄₃₂"	3¹⁷⁄₃₂"	3⁹⁄₁₆"	4⁵⁄₃₂"
45°	2⁷⁄₃₂"	3¹⁹⁄₃₂"	4⁵⁄₁₆"	4⅝"	5¹⁷⁄₃₂"
60°	2²³⁄₃₂"	4⅜"	5¹³⁄₃₂"	5¾"	7"

EMT TAKE-UP FOR HAND BENDERS

Hand Bender	Take-up
½"	5"
¾"	6"
1"	8"
1¼"	12"

GAIN AND RADIUS

Conduit Size	Gain	Radius
½"	2⅝"	4"
¾"	3¼"	5"
1"	4"	6"
1¼"	5⅝"	8"

SUPPORT SPACING FOR RIGID METAL CONDUIT

Conduit Size	Maximum Distance between Supports
½" to ¾"	10'
1"	12'
1¼" to 1½"	14'
2" to 2½"	16'
3" and larger	20'

SUPPORT SPACING FOR EMT AND METAL CONDUIT

All sizes: Every 10', plus within 36" of every box or other termination.

SUPPORT SPACING FOR RIGID PVC

Conduit Size	Maximum Distance between Supports
½" to 1"	3'
1¼" to 1"	5'
2½" to 3"	6'
3½" to 5"	7'
6"	8'

SUPPORT SPACING FOR FLEXIBLE CONDUITS

All sizes: Every 4'6", plus within 12" of every box or other termination.

EXPANSION CHARACTERISTICS OF PVC RIGID NONMETALLIC CONDUIT

Temperature Change in Degrees F	Length Change in Inches per 100 ft	Temperature Change in Degrees F	Length Change in Inches per 100 ft
10°	0.41	110°	4.46
20°	0.81	120°	4.87
30°	1.22	130°	5.27
40°	1.62	140°	5.68
50°	2.03	150°	6.08
60°	2.43	160°	6.49
70°	2.84	170°	6.90
80°	3.24	180°	7.30
90°	3.65	190°	7.71
100°	4.06	200°	8.11

CONDUIT AND TUBING—ALLOWABLE AREA DIMENSIONS FOR WIRE COMBINATIONS

Trade Size Inches	Trade I.D. Inches	100% Total Area Sq. In.	31% 2 Wires Sq. In.	40% Over 2 Wires Sq. In.	53% 1 Wire Sq. In.
RIGID METAL CONDUIT					
½	0.632	0.314	0.097	0.125	0.166
¾	0.836	0.549	0.170	0.220	0.291
1	1.063	0.887	0.275	0.355	0.470
1¼	1.394	1.526	0.473	0.610	0.809
1½	1.624	2.071	0.642	0.829	1.098
2	2.083	3.408	1.056	1.363	1.806
2½	2.489	4.866	1.508	1.946	2.579
3	3.090	7.499	2.325	3.000	3.974
3½	3.570	10.010	3.103	4.004	5.305
4	4.050	12.882	3.994	5.153	6.828
5	5.073	20.212	6.266	8.085	10.713
6	6.093	29.158	9.039	11.663	15.454
LIQUIDTIGHT FLEXIBLE METAL CONDUIT					
⅜	0.494	0.192	0.059	0.077	0.102
½	0.632	0.314	0.097	0.125	0.166
¾	0.830	0.541	0.168	0.216	0.287
1	1.054	0.873	0.270	0.349	0.462
1¼	1.395	1.528	0.474	0.611	0.810
1½	1.588	1.981	0.614	0.792	1.050
2	2.033	3.246	1.006	1.298	1.720
2½	2.493	4.881	1.513	1.953	2.587
3	3.085	7.475	2.317	2.990	3.962
3½	3.520	9.731	3.017	3.893	5.158
4	4.020	12.692	3.935	5.077	6.727
LIQUIDTIGHT FLEXIBLE NONMETALLIC CONDUIT (TYPE LFNC-A)					
⅜	0.495	0.192	0.060	0.077	0.102
½	0.630	0.312	0.097	0.125	0.165
¾	0.825	0.535	0.166	0.214	0.283
1	1.043	0.854	0.265	0.342	0.453
1¼	1.383	1.502	0.466	0.601	0.796
1½	1.603	2.018	0.626	0.807	1.070
2	2.063	3.343	1.036	1.337	1.772
LIQUIDTIGHT FLEXIBLE NONMETALLIC CONDUIT (TYPE LFNC-B)					
⅜	0.494	0.192	0.059	0.077	0.102
½	0.632	0.314	0.097	0.125	0.166
¾	0.830	0.541	0.168	0.216	0.287
1	1.054	0.873	0.270	0.349	0.462
1¼	1.395	1.528	0.474	0.611	0.810
1½	1.588	1.981	0.614	0.792	1.050
2	2.033	3.246	1.006	1.298	1.720

CONDUIT AND TUBING—ALLOWABLE AREA DIMENSIONS FOR WIRE COMBINATIONS

Trade Size Inches	Trade I.D. Inches	100% Total Area Sq. In.	31% 2 Wires Sq. In.	40% Over 2 Wires Sq. In.	53% 1 Wire Sq. In.
ELECTRICAL NONMETALLIC TUBING					
½	0.560	0.246	0.076	0.099	0.131
¾	0.760	0.454	0.141	0.181	0.240
1	1.000	0.785	0.243	0.314	0.416
1¼	1.340	1.410	0.437	0.564	0.747
1¼	1.570	1.936	0.600	0.774	1.026
2	2.020	3.205	0.993	1.282	1.699
ELECTRICAL METALLIC TUBING					
½	0.622	0.304	0.094	0.122	0.161
¾	0.824	0.533	0.165	0.213	0.283
1	1.049	0.864	0.268	0.346	0.458
1¼	1.380	1.496	0.464	0.598	0.793
1½	1.610	2.036	0.631	0.814	1.079
2	2.067	3.356	1.040	1.342	1.778
2½	2.731	5.858	1.816	2.343	3.105
3	3.356	8.846	2.742	3.538	4.688
3½	3.834	11.545	3.579	4.618	6.119
4	4.334	14.753	4.573	5.901	7.819
INTERMEDIATE METAL CONDUIT					
½	0.660	0.342	0.106	0.137	0.181
¾	0.864	0.586	0.182	0.235	0.311
1	1.105	0.959	0.297	0.384	0.508
1¼	1.448	1.647	0.510	0.659	0.873
1½	1.683	2.225	0.690	0.890	1.179
2	2.150	3.630	1.125	1.452	1.924
2½	2.557	5.135	1.592	2.054	2.722
3	3.176	7.922	2.456	3.169	4.199
3½	3.671	10.584	3.281	4.234	5.610
4	4.166	13.631	4.226	5.452	7.224
FLEXIBLE METAL CONDUIT					
⅜	0.384	0.116	0.036	0.046	0.061
½	0.635	0.317	0.098	0.127	0.168
¾	0.824	0.533	0.165	0.213	0.283
1	1.020	0.817	0.253	0.327	0.433
1¼	1.275	1.277	0.396	0.511	0.677
1½	1.538	1.858	0.576	0.743	0.985
2	2.040	3.269	1.013	1.307	1.732
2½	2.500	4.909	1.522	1.963	2.602
3	3.000	7.069	2.191	2.827	3.746
3½	3.500	9.621	2.983	3.848	5.099
4	4.000	12.566	3.896	5.027	6.660

CONDUIT AND TUBING—ALLOWABLE AREA
DIMENSIONS FOR WIRE COMBINATIONS

Trade Size Inches	Trade I.D. Inches	100% Total Area Sq. In.	31% 2 Wires Sq. In.	40% Over 2 Wires Sq. In.	53% 1 Wire Sq. In.
PVC CONDUIT—TYPE EB					
2	2.221	3.874	1.201	1.550	2.053
3	3.330	8.709	2.700	3.484	4.616
3½	3.804	11.365	3.523	4.546	6.023
4	4.289	14.448	4.479	5.779	7.657
5	5.316	22.195	6.881	8.878	11.763
6	6.336	31.530	9.774	12.612	16.711
RIGID PVC CONDUIT—TYPE A					
½	0.700	0.385	0.119	0.154	0.204
¾	0.910	0.650	0.202	0.260	0.345
1	1.175	1.084	0.336	0.434	0.575
1¼	1.500	1.767	0.548	0.707	0.937
1½	1.720	2.324	0.720	0.929	1.231
2	2.155	3.647	1.131	1.459	1.933
2½	2.635	5.453	1.690	2.181	2.890
3	3.230	8.194	2.540	3.278	4.343
3½	3.690	10.694	3.315	4.278	5.668
4	4.180	13.723	4.254	5.489	7.273
RIGID PVC SCHEDULE 40 CONDUIT (HDPE)					
½	0.602	0.285	0.088	0.114	0.151
¾	0.804	0.508	0.157	0.203	0.269
1	1.029	0.832	0.258	0.333	0.441
1¼	1.360	1.453	0.450	0.581	0.770
1½	1.590	1.986	0.616	0.794	1.052
2	2.047	3.291	1.020	1.316	1.744
2½	2.445	4.695	1.455	1.878	2.488
3	3.042	7.268	2.253	2.907	3.852
3½	3.521	9.737	3.018	3.895	5.161
4	3.998	12.554	3.892	5.022	6.654
5	5.016	19.761	6.126	7.904	10.473
6	6.031	28.567	8.856	11.427	15.141
RIGID PVC SCHEDULE 80 CONDUIT					
½	0.526	0.217	0.067	0.087	0.115
¾	0.722	0.409	0.127	0.164	0.217
1	0.936	0.688	0.213	0.275	0.365
1¼	1.255	1.237	0.383	0.495	0.656
1½	1.476	1.711	0.530	0.684	0.907
2	1.913	2.874	0.891	1.150	1.523
2½	2.290	4.119	1.277	1.647	2.183
3	2.864	6.442	1.997	2.577	3.414
3½	3.326	8.688	2.693	3.475	4.605
4	3.786	11.258	3.490	4.503	5.967
5	4.768	17.855	5.535	7.142	9.463
6	5.709	25.598	7.935	10.239	13.567

DIMENSIONS AND WEIGHTS
OF RIGID STEEL CONDUIT

Nominal or Trade Size of Conduit (in)	Inside Diameter (in)	Outside Diameter (in)	Wall Thickness (in)	Length without Coupling (ft & in)	Minimum Weight of Ten Unit Lengths with Couplings Attached (lb)
¼	0.364	0.540	0.088	9–11½	38.5
⅜	0.493	0.675	0.091	9–11½	51.5
½	0.622	0.840	0.109	9–11¼	79.0
¾	0.824	1.050	0.113	9–11¼	105.0
1	1.049	1.315	0.133	9–11	153.0
1¼	1.380	1.660	0.140	9–11	201.0
1½	1.610	1.900	0.145	9–11	249.0
2	2.067	2.375	0.154	9–11	334.0
2½	2.469	2.875	0.203	9–10½	527.0
3	3.068	3.500	0.216	9–10½	690.0
3½	3.548	4.000	0.226	9–10½	831.0
4	4.026	4.500	0.237	9–10¼	982.0
5	5.047	5.563	0.258	9–10	1,344.0
6	6.065	6.625	0.280	9–10	1,770.0

Note: The tolerances are:
   Length: ± ¼ inch (without coupling)
   Outside Diameter: + ⁄₆₄ inch or –⁄₃₂ inch for the 1½-inch and smaller sizes
      ± 1 percent for the 2-inch and larger sizes
   Wall Thickness: – 12½ percent

DIMENSIONS OF THREADS FOR
RIGID STEEL CONDUIT

Nominal or Trade Size of Conduit (in)	Threads per Inch	Pitch Diameter at End of Thread E_0 (in)	Length of Thread (in)	
		Taper ¾ inch per Foot	Effective L_2	Overall L_4
¼	18	0.4774	0.40	0.59
⅜	18	0.6120	0.41	0.60
½	14	0.7584	0.53	0.78
¾	14	0.9677	0.55	0.79
1	11½	1.2136	0.68	0.98
1¼	11½	1.5571	0.71	1.01
1½	11½	1.7961	0.72	1.03
2	11½	2.2690	0.76	1.06
2½	8	2.7195	1.14	1.57
3	8	3.3406	1.20	1.63
3½	8	3.8375	1.25	1.68
4	8	4.3344	1.30	1.73
5	8	5.3907	1.41	1.84
6	8	6.4461	1.51	1.95

Note: The tolerances are:
   Thread Length (L_4, Column 5): ± 1 thread
   Pitch Diameter (Column 3): ± 1 turn is the maximum variation permitted from the gauging face of the working thread gauges. This is equivalent to ± 1-½ turns from basic dimensions, since a variation of ±½ turn from basic dimensions is permitted in working gauges.

DIMENSIONS AND WEIGHTS OF RIGID STEEL COUPLINGS

Nominal or Trade Size of Conduit (in)	Outside Diameter (in)	Minimum Length (in)	Minimum Weight (lb)
1/4	0.719	1 3/16	0.055
3/8	0.875	1 3/16	0.075
1/2	1.010	1 9/16	0.115
3/4	1.250	1 5/8	0.170
1	1.525	2	0.300
1 1/4	1.869	2 1/16	0.370
1 1/2	2.155	2 1/16	0.515
2	2.650	2 1/8	0.671
2 1/2	3.250	3 1/8	1.675
3	3.870	3 1/4	2.085
3 1/2	4.500	3 3/8	3.400
4	4.875	3 1/2	2.839
5	6.000	3 3/4	4.462
6	7.200	4	7.282

Note: The tolerances are:
  Outside Diameter: −1 percent for the 1 1/4-inch and larger sizes.
             −1/64 inch for sizes smaller than 1 1/4 inch.
  No limit is placed on the plus tolerances given for this dimension.

DIMENSIONS AND WEIGHTS OF RIGID STEEL 90° ELBOWS AND NIPPLES

Nominal or Trade Size of Conduit (in)	Elbows		Nipples	
	Minimum Radius to Center of Conduit (in)	Minimum Straight Length Ls at Each End (in)	A	B
1/4	—	—	—	—
3/8	—	—	—	—
1/2	4	1 1/2	0.065	2
3/4	4 1/2	1 1/2	0.086	4
1	5 3/4	1 7/8	0.125	9
1 1/4	7 1/4	2	0.164	10
1 1/2	8 1/4	2	0.202	11
2	9 1/2	2	0.269	14
2 1/2	10 1/2	3	0.430	60
3	13	3 1/8	0.561	70
3 1/2	15	3 1/4	0.663	90
4	16	3 3/8	0.786	115
5	24	3 5/8	1.060	170
6	30	3 3/4	1.410	200

Each lot of 100 nipples shall weigh not less than the number of pounds determined by the formula:
$$W = 100\,LA - B$$

Where W = weight of 100 nipples in pounds   L = length of one nipple in inches
    A = weight of nipple per inch in pounds
    B = weight in pounds lost in threading 100 nipples

DIMENSIONS AND WEIGHTS OF EMT

Nominal or Trade Size of Tubing (in)	Outside Diameter (in)	Minimum Wall Thickness (in)	Length (ft)	Minimum Weight per 100 Feet (lb)
⅜	0.577	0.040	10	23
½	0.706	0.040	10	28.5
¾	0.922	0.046	10	43.5
1	1.163	0.054	10	64
1¼	1.510	0.061	10	95
1½	1.740	0.061	10	110
2	2.197	0.061	10	140

Note: The tolerances are:
Length: ± ¼ inch
Outside Diameter: ± 0.005 inch
Wall Thickness: + 18 percent

DIMENSIONS OF EMT 90° ELBOWS

Nominal or Trade Size of Tubing (in)	Minimum Radius to Center of Tubing (in)	Minimum Straight Length Ls at Each End (in)
½	4	1½
¾	4½	1½
1	5¾	1⅞
1¼	7¼	2
1½	8¼	2
2	9½	2

CONDUIT MASTER BUNDLES IN TOTAL LENGTH AND WEIGHT

R.M.C.			E.M.T.		
Size (in)	Length (ft)	Weight (lb)	Size (in)	Length (ft)	Weight (lb)
½	2,500	2,050	½	7,000	2,100
¾	2,000	2,180	¾	5,000	2,300
1	1,250	2,013	1	3,000	2,010
1¼	900	1,962	1¼	2,000	2,020
1½	800	2,104	1½	1,500	1,740
2	600	2,100	2	1,200	1,776
2½	370	2,068	2½	610	1,318
3	300	2,181	3	510	1,341
3½	250	2,200	3½	370	1,291
4	200	2,060	4	300	1,179
5	150	2,100			
6	100	1,840			

I.M.C.			Schedule 40 PVC		
Size (in)	Length (ft)	Weight (lb)	Size (in)	Length (ft)	Weight (lb)
½	3,500	2,170	½	6,000	972
¾	2,500	2,100	¾	4,400	951
1	1,700	2,023	1	3,600	1,152
1¼	1,350	2,133	1¼	3,300	1,433
1½	1,100	2,134	1½	2,250	1,170
2	800	2,048	2	1,400	979
2½	370	1,632	2½	930	1,031
3	300	1,629	3	880	1,276
3½	240	1,510	3½	630	1,099
4	240	1,680	4	570	1,178
			5	380	1,065
			6	260	946

NONMETALLIC-SHEATHED CABLE
(TYPES NM AND NMC)

- Covered by Article 334 of the NEC®.
- Referred to as Romex of sizes #14 to #2 AWG with copper conductors and an equipment grounding conductor of proper size.
- The grounding conductor may be bare or covered with green insulation, or green insulation with one or more yellow stripes.
- NM may be used for both exposed and concealed work in normally dry locations.
- NM and NMC may not be used as service-entrance cable, embedded in poured cement, concrete, or aggregate. See the NEC for structures more than three floors above grade.
- NMC may be used for both exposed and concealed work in dry, moist, damp, or corrosive locations.
- NM and NMC cables must be secured within 12 inches of every cabinet, box, or fitting and at least every 4½ feet elsewhere.
- NM and NMC cables can be damaged due to sharp bending, being pulled through bored holes, or by driving the staples too forcefully.

UNDERGROUND FEEDER AND
BRANCH-CIRCUIT CABLE (TYPE UF)

- Covered by Article 340 of the NEC®.
- Sizes #14 to 4/0 AWG with copper conductors and an equipment grounding conductor of proper size.
- The grounding conductor may be bare or covered with green insulation, or green insulation with one or more yellow stripes.
- Type UF cable may be buried directly in the earth at a minimum depth of 24 inches. This may be reduced to 18 inches beneath a 2-inch concrete pad, or encased in a metal raceway, pipe, or other suitable protection.
- UF cable may not be used as service-entrance cable, embedded in poured cement, concrete, or aggregate, or exposed to sunlight.

ARMORED CABLE (TYPE AC)

- Covered by Article 320 of the NEC®.
- Referred to as BX and contains a bonding strip.
- AC cable may also contain an equipment grounding conductor of proper size that may be bare or covered with green insulation, or green insulation with one or more yellow stripes.
- May be used for both exposed and concealed work in normally dry locations and may be embedded in plaster finish on brick or other masonry, except in damp or wet locations.
- May be run or fished in the air voids of masonry block or tile walls, except where such walls are exposed or subject to excessive moisture or dampness or are below grade line.
- May not be installed in damp or wet locations, or where exposed to damage.
- AC cables must be secured within 12 inches of every cabinet, box, or fitting and at least every 4½ feet elsewhere.
- AC cables can be damaged due to sharp bending.
- At each termination, a fiber bushing must be inserted between the armored sheath and the conductors. The grounding strip must be bent back over the fiber bushing and in intimate contact with the external armor and the cable clamp.

SERVICE-ENTRANCE CABLE
(TYPES SE AND USE)

- Covered by Article 338 of the NEC®.
- Used for service entrances of 100 amps.
- May be used for electric ranges and may be called Range cable.
- Cable will also contain an equipment grounding conductor of proper size that may be bare or covered with green insulation, or green insulation with one or more yellow stripes.
- SE cables must be secured within 12 inches of every cabinet, box, or fitting and at least every 4½ feet elsewhere.

UNDERGROUND INSTALLATION REQUIREMENTS

Conductors and cables shall be protected to 8' above finished grade

Finished grade

Warning tape

Raceway protected below grade

Minimum 12" above installation

Heavy rocks or sharp corrosive material prohibited as backfill

Finished grade

Trench

Raceway

UNDERGROUND
INSTALLATION REQUIREMENTS *(cont.)*

Enclosure

Raceway must be sealed to prevent moisture

Finished grade

Bushing required at end of conduit underground with conductors

Underground conduit

"S" loop permitted for ground movement

Finished grade

Cable — Splice

Splice box not required

MINIMUM COVER REQUIREMENTS FOR UNDERGROUND INSTALLATIONS

Type of circuit or wiring method (0 to 600 V, nominal)

Location of circuit or wiring method	Direct burial cables or conductors	Rigid metal conduit or intermediate metal conduit	Nonmetallic raceways listed for direct burial without concrete encasement or other approved raceways	Residential branch circuits rated 120 V or less with GFCI protection and maximum overcurrent protection of 20 amperes	Circuits for control of irrigation and landscape lighting limited to not more than 30 V and installed with type UF or in other identified cables or raceways
All locations not specified in these charts	24"	6"	18"	12"	6"
In trench below 2" thick concrete or equivalent	18"	6"	12"	6"	6"
Under buildings	In raceway only	—	—	In raceway only	In raceway only
Under minimum of 4" thick concrete exterior slab with no traffic and slab extending no less than 6" beyond the installation	18"	4"	4"	6" (Direct burial) 4" (In raceway)	6" (Direct burial) 4" (In raceway)
Under alleys, highways, roads, driveways, streets, and parking lots	24"	24"	24"	24"	24"

MINIMUM COVER REQUIREMENTS FOR UNDERGROUND INSTALLATIONS *(cont.)*

Type of circuit or wiring method (0 to 600 V, nominal)

Location of circuit or wiring method	Direct burial cables or conductors	Rigid metal conduit or intermediate metal conduit	Nonmetallic raceways listed for direct burial without concrete encasement or other approved raceways	Residential branch circuits rated 120 V or less with GFCI protection and maximum over-current protection of 20 amperes	Circuits for control of irrigation and landscape lighting limited to not more than 30 V and installed with type UF or in other identified cables or raceways
Under family dwelling driveways. Outdoor parking areas used for dwelling-related purposes only	18"	18"	18"	12"	18"
In or under airport runways, including areas where the public is prohibited	18"	18"	18"	18"	18"

Notes: Cover is defined as the shortest distance between a point on the top service of any direct-buried conductor, cable, conduit, or other raceway and the top surface of finished grade, concrete, etc.

Raceways approved for burial only where concrete encased shall require a 2" concrete envelope.

Lesser depths are permitted where cables/conductors rise for terminations, splices, or equipment.

Where solid rock prevents compliance with the cover depths specified in this table, the wiring must be installed in raceways permitted for direct burial. The raceways shall be covered by a minimum of 2" of concrete extending down to the solid rock.

MINIMUM BURIAL DEPTH REQUIREMENTS ABOVE 600 V

Nominal Voltage to Ground	Directly-Buried Cables	Rigid Non metallic Conduit	Rigid or Intermediate Metal Conduit	Under 4" Minimum Building Slab	Cables Under or Near Airport Runways, Restricted Areas	Under Roads & Parking Lots
600 to 22,000 V	30"	18"	6"	4"	18"	24"
22,000 to 40,000 V	36"	24"	6"	4"	18"	24"
Above 40,000 V	42"	30"	6"	4"	18"	24"
See NEC® Table 300.50 for complete details.						

CHAPTER 3
Ampacity, Box Fill, Branch Circuits, Conductors, Enclosures, and Raceways

AMPACITY OF LAMP AND EXTENSION CORDS — TYPES S, SJ, SJT, SP, SPT, AND ST

Wire Size (AWG)	Current in Amps		
	4 Conductors	3 Conductors	2 Conductors
18	6	7	10
16	8	10	13
14	12	15	18
12	16	20	25
10	20	25	30

VERTICAL CONDUCTOR SUPPORTS

Size of Wire (AWG, Circular Mil)	Conductor Type	
	Copper	Aluminum or Copper-Clad Aluminum
18 through 8	100 feet	100 feet
6 through 1/0	100 feet	200 feet
2/0 through 4/0	80 feet	180 feet
Over 4/0 through 350	60 feet	135 feet
Over 350 through 500	50 feet	120 feet
Over 500 through 750	40 feet	95 feet
Over 750	35 feet	85 feet

For SI units: 1 foot = 0.3048 meter.

CONDUCTOR COLOR CODE

Grounded Conductor
- White
- Gray
- Three continuous white stripes

Ungrounded Conductor
- Any color other than white, gray, or green

Equipment Grounding Conductor
- Green with one or more yellow stripes
- Bare

POWER WIRING COLOR CODE

120/240 V		277/480 V	
Black	Phase 1	Brown	Phase 1
Red	Phase 2	Orange	Phase 2
Blue	Phase 3	Yellow	Phase 3
Gray, white, or with 3 white stripes	Neutral	Gray or with 3 white stripes	Neutral
Green	Ground	Green with yellow stripe	Ground

POWER TRANSFORMER COLOR CODE

Wire Color	Transformer Circuit Type
Black	If a transformer does not have a tapped primary, both leads are black.
Black	If a transformer does have a tapped primary, the black is the common lead.
Black and Yellow	Tap for a tapped primary
Black and Red	End for a tapped primary

GROUPED CONDUCTORS

**277/480 V, 3φ,
4-wire circuit**

All phase conductors, grounded conductors, and equipment ground conductors shall be grouped together so as not to cause induction heating of metal raceways and enclosures.

PARALLELED CONDUCTORS

**Three parallel 4/0
THHN copper conductors
per phase**

Requirements:

**Same size, minimum
1/0 AWG**

Same materials

Same length

Same insulation type

Same attachments

ELECTRICAL CABLE CLASS RATINGS

An electrical cable is rated according to the following parameters: the number of wires, the wire size, the type of insulation, and the moisture condition of the environment of the wire. Therefore, an electrical cable designated 10/2 with ground—type UF—600 V-(UL) meets the specifications below:

- The "10" relates to wire size—10 gauge wire.
- 2 defines an electrical cable with two wires.
- The word "ground" indicates the cable has a third wire to be connected to ground.
- The term "type UF" means the insulation type has an acceptable moisture rating and is an underground feeder type cable.
- "600 V" defines the cable as being rated at 600 volts maximum.
- "(UL)" means the cable has certification from Underwriters Laboratory.

CABLE INSULATION MOISTURE RATINGS

Dry	Indoor above ground level; moisture usually not encountered.
Damp	Indoor below ground level (basement); locations are partially protected; moisture level is moderate.
Wet	Locations affected by weather (outside); concrete slabs, underground, etc.; water saturation likely.

CONDUCTOR PREFIX CODES

B	Outer braid		O	Neoprene jacket
F	Fixture wire		R	Rubber covering
FEP	Fluorinated ethylene propylene, used only in locations hotter than 90°C		S	Appliance cord
			SP	Lamp cord, rubber
			SPT	Lamp cord, plastic
H	Load temp up to 75°C		T	Load temp up to 60°C
HH	Load temp up to 90°C		W	Wet use only
L	Seamless lead jacket		X	Moisture and heat resistant
M	Machine tool wire		-Z	Rated for both wet and dry use (THWN-Z)
N	Resistant to oil and gas			

CONDUCTOR APPLICATIONS

Type	Max Temp	Application	Insulation	Outer Covering
FEP or FEPB	90°C (194°F) 200°C (392°F)	Dry and damp locations Dry locations – Special Apps	Fluorinated ethylene propylene	None or glass braid
MI	90°C (194°F) 250°C (482°F)	Dry and wet locations Special Apps	Magnesium oxide	Copper or alloy steel
MTW	60°C (140°F) 90°C (194°F)	Machine tool wiring – wet locations Machine tool wiring – dry locations	Flame-retardant, moisture-, heat-, and oil-resistant thermoplastic	None or nylon jacket
PAPER	85°C (185°F)	Underground service conductors	Paper	Lead sheath
PFA	90°C (194°F) 200°C (392°F)	Dry and damp locations Dry locations—Special Apps	Perfluoroalkoxy	None
PFAH	250°C (482°F)	Dry locations only	Perfluoroalkoxy	None
RHH	90°C (194°F)	Dry and damp locations		Moisture resistant, flame-retardant nonmetallic
RHW	75°C (167°F)	Dry and wet locations	Flame-retardant, moisture-resistant thermoset	Moisture-resistant, flame-retardant nonmetallic

CONDUCTOR APPLICATIONS (cont.)

Type	Max Temp	Application	Insulation	Outer Covering
RHW-2	90°C (194°F)	Dry and wet locations	Flame-retardant, moisture-resistant thermoset	Moisture-resistant, flame-retardant, nonmetallic
SA	90°C (194°F) 200°C (392°F)	Dry and damp locations Special Apps	Silicone rubber	Glass or braid material
SIS	90°C (194°F)	Switchboard wiring	Flame-retardant thermostat	None
TBS	90°C (194°F)	Switchboard wiring	Thermoplastic	Flame-retardant, nonmetallic
TFE	250°C (482°F)	Dry locations only	Extruded polytetrafluoroethylene	None
THHN	90°C (194°F)	Dry and damp locations	Flame-retardant, heat-resistant thermoplastic	Nylon jacket
THHW	75°C (167°F) 90°C (194°F)	Wet locations Dry locations	Flame-retardant, moisture- and heat-resistant thermoplastic	None
THW THW-2	75°C (167°F) 90°C (194°F)	Dry and wet locations Special Apps	Flame-retardant, moisture- and heat-resistant thermoplastic	None
THWN THWN-2	75°C (167°F) 90°C (194°F)	Dry and wet locations Special Apps	Flame-retardant, moisture- and heat-resistant thermoplastic	Nylon jacket
TW	60°C (140°F)	Dry and wet locations	Flame-retardant, moisture- and heat-resistant thermoplastic	None
UF	60°C (140°F) 75°C (167°F)	Refer to NEC®	Moisture- and heat-resistant	Integral with insulation

CONDUCTOR APPLICATIONS (cont.)

Type	Max Temp	Application	Insulation	Outer Covering
USE USE-2	75°C (167°F) 90°C (194°F)	Refer to NEC® Dry and wet locations	Heat- and moisture-resistant	Moisture-resistant nonmetallic
XHH	90°C (194°F)	Dry and damp locations	Flame-retardant thermoplastic	None
XHHW	90°C (194°F) 75°C (167°F)	Dry and damp locations Wet locations	Flame-retardant, moisture-resistant thermoset	None
XHHW-2	90°C (194°F)	Dry and wet locations	Flame-retardant, moisture-resistant thermoset	None
Z	90°C (194°F) 150°C (302°F)	Dry and damp locations Dry locations – Special Apps	Modified ethylene Tetrafluoro ethylene	None
ZW	75°C (167°F) 90°C (194°F) 150°C (302°F)	Wet locations Dry and damp locations Dry locations – Special Apps	Modified ethylene Tetrafluoro ethylene	None
ZW-2	90°C (194°F)	Dry and wet locations	Modified ethylene Tetrafluoro ethylene	None

ENCLOSURE TYPES

Type	Use	Protection Against	UL Tests	Comments
1	Indoor	Incidental contact, falling dirt	Rod entry, rust resistance	—
3	Outdoor	Windblown dust, rain, sleet, and ice on enclosure	Rain, external icing, dust, and rust resistance	No protection against internal condensation or internal icing
3R	Outdoor	Falling rain and ice on enclosure	Rod entry, rain, external icing, and rust resistance	No protection against dust, internal condensation, or internal icing
4	Indoor/outdoor	Windblown dust and rain, splashing water, hose-directed water, and ice on enclosure	Hosedown, external icing, and rust resistance	No protection against internal condensation or internal icing
4X	Indoor/outdoor	Corrosion, windblown dust and rain, splashing water, hose-directed water, and ice on enclosure	Hosedown, external icing, and corrosion resistance	No protection against internal condensation or internal icing
6	Indoor/outdoor	Occasional temporary submersion at a limited depth	—	—

ENCLOSURE TYPES (cont.)

Type	Use	Protection Against	UL Tests	Comments
6P	Indoor/outdoor	Prolonged submersion at a limited depth	—	—
7	Indoor locations classified as Class I, Groups A, B, C, or D, as defined in the NEC®	Withstand and contain an internal explosion of specified gases, contain an explosion sufficiently so an explosive gas–air mixture in the atmosphere is not ignited	Explosion, hydrostatic, and temperature	Enclosed heat-generating devices shall not cause external surfaces to reach temperatures capable of igniting explosive gas–air mixtures in the atmosphere
9	Indoor locations classified as Class II, Groups E, F, or G, as defined in the NEC®	Dust	Dust penetration, temperature, and gasket aging	Enclosed heat-generating devices shall not cause external surfaces to reach temperatures capable of igniting explosive gas–air mixtures in the atmosphere
12	Indoor	Dust, falling dirt, and dripping noncorrosive liquids	Drip, dust, and rust resistance	No protection against internal condensation
13	Indoor	Dust, spraying water, oil, and noncorrosive coolant	Oil explosion and rust resistance	No protection against internal condensation

CLASSIFICATIONS OF HAZARDOUS LOCATIONS

Classes	Likelihood that a flammable or combustible concentration is present
I	Sufficient quantities of flammable gases and vapors present in air to cause an explosion or ignite hazardous materials.
II	Sufficient quantities of combustible dust are present in air to cause an explosion or ignite hazardous materials.
III	Easily-ignitable fibers or flyings are present in air, but not in a sufficient quantity to cause an explosion or ignite hazardous materials.

Divisions	Location containing hazardous substances
1	Hazardous location in which hazardous substance is **normally present** in air in sufficient quantities to cause an explosion or ignite hazardous materials.
2	Hazardous location in which hazardous substance is **not normally present** in air in sufficient quantities to cause an explosion or ignite hazardous materials.

Groups	Atmosphere containing flammable gases or vapors or combustible dust	
Class I	Class II	Class III
A	E	
B	F	None
C	G	
D		

DIVISION 1 EXAMPLES

Class I

- Spray booth interiors
- Areas adjacent to spraying or painting operations using volatile flammable solvents
- Open tanks or vats of volatile flammable liquids
- Drying or evaporation rooms for flammable solvents
- Areas where fats and oil extraction equipment using flammable solvents are operated
- Cleaning and dyeing plant rooms that use flammable liquids
- Gas generator rooms
- Pump rooms for flammable gases or volatile flammable liquids that do not contain adequate ventilation
- Refrigeration or freezer interiors that store flammable materials
- All other locations where sufficient ignitable quantities of flammable gases or vapors are likely to occur during routine operations

Class II

- Grain and grain products
- Pulverized sugar and cocoa
- Dried egg and milk powders
- Pulverized spices
- Starch and pastes
- Potato and woodflour
- Oil meal from beans and seeds
- Dried hay
- Any other organic materials that may produce combustible dusts during their use or handling

Class III

- Portions of rayon, cotton, or other textile mills
- Manufacturing and processing plants for combustible fibers, cotton gins, and cotton seed mills
- Flax-processing plants
- Clothing-manufacturing plants
- Woodworking plants
- Other establishments involving similar hazardous processes or conditions

SIZES OF PANELBOARDS

Single-Phase — 3-Wire Systems

40 A	100 A	150 A	225 A	400 A
70 A	125 A	200 A	300 A	600 A

Three-Phase — 4-Wire Systems

60 A	150 A	225 A	400 A
125 A	200 A	300 A	600 A

SIZES OF GUTTERS AND WIREWAYS

2½" × 2½"	6" × 6"	10" × 10"
4" × 4"	8" × 8"	

These sizes are available in 12", 24", 36", 48", and 60" lengths.

SIZES OF DISCONNECTS

30 A	200 A	800 A	1600 A
60 A	400 A	1200 A	1800 A
100 A	600 A	1400 A	

SIZES OF PULL BOXES AND JUNCTION BOXES

4" × 4" × 4"	10" × 8" × 4"	12" × 12" × 6"
6" × 4" × 4"	10" × 8" × 6"	12" × 12" × 8"
6" × 6" × 4"	10" × 10" × 4"	15" × 12" × 4"
6" × 6" × 6"	10" × 10" × 6"	15" × 12" × 6"
8" × 6" × 4"	10" × 10" × 8"	18" × 12" × 4"
8" × 6" × 6"	12" × 8" × 4"	18" × 12" × 6"
8" × 6" × 8"	12" × 8" × 6"	18" × 18" × 4"
8" × 8" × 4"	12" × 10" × 4"	18" × 18" × 6"
8" × 8" × 6"	12" × 10" × 6"	24" × 18" × 6"
8" × 8" × 8"	12" × 12" × 4"	24" × 24" × 6"
		24" × 24" × 8"

BUSWAY OR BUSDUCT

1φ	3φ
225 A	225 A
400 A	400 A
600 A	600 A
800 A	800 A
1,000 A	1,000 A
1,200 A	1,200 A
1,350 A	1,350 A
1,600 A	1,600 A
2,000 A	2,000 A
2,500 A	2,500 A
3,000 A	3,000 A
4,000 A	4,000 A
5,000 A	5,000 A

CBs AND FUSES

15	70	225	800
20	80	250	1,000
25	90	300	1,200
30	100	350	1,600
35	110	400	2,000
40	125	450	2,500
45	150	500	3,000
50	175	600	4,000
60	200	700	5,000
			6,000

For fuses only, additional standard sizes are 1, 3, 6, and 10. All sizes are rated in amps.

SWITCHBOARDS OR SWITCHGEARS

1φ	3φ
200 A	400 A
400 A	600 A
600 A	800 A
800 A	1,200 A
1,200 A	1,600 A
1,600 A	2,000 A
2,000 A	2,500 A
2,500 A	3,000 A
3,000 A	4,000 A
4,000 A	

AREA DIMENSIONS OF INSULATED CONDUCTORS

Type	Size	Inch²	Type	Size	Inch²
RFH-2,	18	0.0145	RHH*, RHW*, RHW-2*	14	0.0209
FFH-2	16	0.0172	ABOVE AND XF, XFF	12	0.0260
RHW-2, RHH,	14	0.0293	RHH*, RHW*, RHW-2*,	10	0.0333
RHW	12	0.0353	XF, XFF		
	10	0.0437	RHH*, RHW*, RHW-2*	8	0.0556
	8	0.0835			
	6	0.1041	TW, THW,	6	0.0726
	4	0.1333	THHW,	4	0.0973
	3	0.1521	THW-2,	3	0.1134
	2	0.1750	RHH*,	2	0.1333
	1	0.2660	RHW*,	1	0.1901
	1/0	0.3039	RHW-2*	1/0	0.2223
	2/0	0.3505		2/0	0.2624
	3/0	0.4072		3/0	0.3117
	4/0	0.4754		4/0	0.3718
	250	0.6291		250	0.4596
	300	0.7088		300	0.5281
	350	0.7870		350	0.5958
	400	0.8626		400	0.6619
	500	1.0082		500	0.7901
	600	1.2135		600	0.9729
	700	1.3561		700	1.1010
	750	1.4272		750	1.1652
	800	1.4957		800	1.2272
	900	1.6377		900	1.3561
	1,000	1.7719		1,000	1.4784
	1,250	2.3479		1,250	1.8602
	1,500	2.6938		1,500	2.1695
	1,750	3.0357		1,750	2.4773
	2,000	3.3719		2,000	2.7818
SF-2, SFF-2	18	0.0115			
	16	0.0139	TFN,	18	0.0055
	14	0.0172	TFFN	16	0.0072
SF-1, SFF-1	18	0.0065	THHN,	14	0.0097
RFH-1, XF, XFF	18	0.0080	THWN,	12	0.0133
TF, TFF, XF, XFF	16	0.0109	THWN-2	10	0.0211
TW, XF, XFF, THHW,	14	0.0139		8	0.0366
THW, THW-2				6	0.0507
TW, THHW,	12	0.0181		4	0.0824
THW, THW-2	10	0.0243		3	0.0973
	8	0.0437		2	0.1158
				1	0.1562

*types minus outer covering

AREA DIMENSIONS OF INSULATED CONDUCTORS (cont.)

Type	Size	Inch^2	Type	Size	Inch^2
THHN,	1/0	0.1855	XHHW, ZW,	14	0.0139
THWN,	2/0	0.2223	XHHW-2,	12	0.0181
THWN-2	3/0	0.2679	XHH	10	0.0243
(cont.)	4/0	0.3237		8	0.0437
	250	0.3970		6	0.0590
	300	0.4608		4	0.0814
	350	0.5242		3	0.0962
	400	0.5863		2	0.1146
	500	0.7073	XHHW,	1	0.1534
	600	0.8676	XHHW-2,	1/0	0.1825
	700	0.9887	XHH	2/0	0.2190
	750	1.0496		3/0	0.2642
	800	1.1085		4/0	0.3197
	900	1.2311		250	0.3904
	1,000	1.3478		300	0.4536
PF, PGFF, PGF, PFF,	18	0.0058		350	0.5166
PTF, PAF, PTFF, PAFF	16	0.0075		400	0.5782
PF, PGFF, PGF, PFF,	14	0.0100		500	0.6984
PTF, PAF, PTFF, PAFF,				600	0.8709
TFE, FEP, PFA,				700	0.9923
FEPB, PFAH				750	1.0532
TFE, FEP,	12	0.0137		800	1.1122
PFA, FEPB,	10	0.0191		900	1.2351
PFAH	8	0.0333		1,000	1.3519
	6	0.0468		1,250	1.7180
	4	0.0670		1,500	2.0157
	3	0.0804		1,750	2.3127
	2	0.0973		2,000	2.6073
TFE, PFAH	1	0.1399	KF-2,	18	0.0031
TFE,	1/0	0.1676	KFF-2	16	0.0044
PFA,	2/0	0.2027		14	0.0064
PFAH, Z	3/0	0.2463		12	0.0093
	4/0	0.3000		10	0.0139
ZF, ZFF	18	0.0045	KF-1,	18	0.0026
	16	0.0061	KFF-1	16	0.0037
Z, ZF, ZFF	14	0.0083		14	0.0055
Z	12	0.0117		12	0.0083
	10	0.0191		10	0.0127
	8	0.0302			
	6	0.0430			
	4	0.0625			
	3	0.0855			
	2	0.1029			
	1	0.1269			

JUNCTION BOX CALCULATIONS

Distance (A) is $6 \times 1" = 6"$ minimum
Distance (B) is $6 \times 2" = 12"$ minimum
Distance (C) is $6 \times 3" = 18"$ minimum
Distance (D) is $6 \times 4" = 24"$ minimum
Distance (E) is $(6 \times 4") + 3" + 2" + 1" = 30"$ minimum
Distance (F) is $(6 \times 4") + 3" + 2" + 1" = 30"$ minimum

CONDUCTOR VOLUME ALLOWANCE

Wire Size (AWG)	Volume Each (in³)	Formula
18	1.50	$V = L \cdot W \cdot D$
16	1.75	Volume =
14	2.00	Length times width
12	2.25	times depth =
10	2.50	cubic inches
8	3.00	
6	5.00	

To find box size needed, add up total volume for all wires to be used. Then use the volume formula; for example, if total volume of all wires is 420 cu in use an $8" \times 10" \times 6"$ (i.e., 480 cu in) box.

BOX FILL

Max Number of Conductors in Outlet, Device, and Junction Boxes

Box Dimension in Inches Trade Size or Type	Min Capacity (in³)	Maximum Number of Conductors per Wire Size (AWG)						
		# 18	# 16	# 14	# 12	# 10	# 8	# 6
4 × 1¼ round or octagonal	12.5	8	7	6	5	5	5	2
4 × 1½ round or octagonal	15.5	10	8	7	6	6	5	3
4 × 2⅛ round or octagonal	21.5	14	12	10	9	8	7	4
4 × 1¼ square	18.0	12	10	9	8	7	6	3
4 × 1½ square	21.0	14	12	10	9	8	7	4
4 × 2⅛ square	30.3	20	17	15	13	12	10	6
4¹¹⁄₁₆ × 1¼ square	25.5	17	14	12	11	10	8	5
4¹¹⁄₁₆ × 1½ square	29.5	19	16	14	13	11	9	5
4¹¹⁄₁₆ × 2⅛ square	42.0	28	24	21	18	16	14	8
3 × 2 × 1½ device	7.5	5	4	3	3	3	2	1
3 × 2 × 2 device	10.0	6	5	5	4	4	3	2
3 × 2 × 2¼ device	10.5	7	6	5	4	4	3	2
3 × 2 × 2½ device	12.5	8	7	6	5	5	4	2
3 × 2 × 2¾ device	14.0	9	8	7	6	5	4	2
3 × 2 × 3½ device	18.0	12	10	9	8	7	6	3
4 × 2⅛ × 1½ device	10.3	6	5	5	4	4	3	2
4 × 2⅛ × 1⅞ device	13.0	8	7	6	5	5	4	2
4 × 2⅛ × 2⅛ device	14.5	9	8	7	6	5	4	2
3¾ × 2 × 2½ masonry box/gang	14.0	9	8	7	6	5	4	2
3¾ × 2 × 3½ masonry box/gang	21.0	14	12	10	9	8	7	2

Where one or more internal cable clamps are present in the box, a single volume allowance (conductor) shall be made based on the largest size conductor in the box.

RIGID METAL CONDUIT—MAXIMUM NUMBER OF CONDUCTORS

Type	Size	RMC Trade Size (in)											
		½	¾	1	1¼	1½	2	2½	3	3½	4	5	6
RHH, RHW, RHW-2	14	4	7	12	21	28	46	66	102	136	176	276	398
	12	3	6	10	17	23	38	55	85	113	146	229	330
	10	3	5	8	14	19	31	44	68	91	118	185	267
	8	1	2	4	7	10	16	23	36	48	61	97	139
	6	1	1	3	6	8	13	18	29	38	49	77	112
	4	1	1	2	4	6	10	14	22	30	38	60	87
	3	1	1	2	4	5	9	12	19	26	34	53	76
	2	1	1	1	3	4	7	11	17	23	29	46	66
	1	0	1	1	1	3	5	7	11	15	19	30	44
	1/0	0	1	1	1	2	4	6	10	13	17	26	38
	2/0	0	1	1	1	2	4	5	8	11	14	23	33
	3/0	0	0	1	1	1	3	4	7	10	12	20	28
	4/0	0	0	1	1	1	3	4	6	8	11	17	24
	250	0	0	0	1	1	1	3	4	6	8	13	18
	300	0	0	0	1	1	1	2	4	5	7	11	16
	350	0	0	0	1	1	1	2	4	5	6	10	15
	400	0	0	0	1	1	1	1	3	4	6	9	13
	500	0	0	0	1	1	1	1	3	4	5	8	11
	600	0	0	0	0	1	1	1	1	2	3	6	9
	700	0	0	0	0	1	1	1	1	1	3	6	8
	750	0	0	0	0	0	1	1	1	1	3	5	8
	800	0	0	0	0	0	1	1	1	1	2	3	7
	1,000	0	0	0	0	0	0	1	1	1	1	3	6
TW	14	9	15	25	44	59	98	140	216	288	370	581	839
	12	7	12	19	33	45	75	107	165	221	284	446	644
	10	5	9	14	25	34	56	80	123	164	212	332	480
	8	3	5	8	14	19	31	44	68	91	118	185	267
RHH*, RHW*, RHW-2*, THHW, THW, THW-2	14	6	10	17	29	39	65	93	143	191	246	387	558
	12	5	8	13	23	32	52	75	115	154	198	311	448
	10	4	6	10	18	25	41	58	90	120	154	242	350
	8	1	4	6	11	15	24	35	54	72	92	145	209
RHH*, RHW*, RHW-2*, TW, THW, THHW, THW-2	6	1	3	5	8	11	18	27	41	55	71	111	160
	4	1	1	3	6	8	14	20	31	41	53	83	120
	3	1	1	3	5	7	12	17	26	35	45	71	103
	2	1	1	2	4	6	10	14	22	30	38	60	87
	1	1	1	1	3	4	7	10	15	21	27	42	61
NOTE:	1/0	0	1	1	2	3	6	8	13	18	23	36	52
(*) Denotes	2/0	0	1	1	2	3	5	7	11	15	19	31	44
types minus	3/0	0	1	1	1	2	4	6	9	13	16	26	37
outer	4/0	0	1	1	1	1	3	5	8	10	14	21	31
covering	250	0	0	1	1	1	3	4	6	8	11	17	25
	300	0	0	1	1	1	2	3	5	7	9	15	22
	350	0	0	0	1	1	1	3	5	6	8	13	19
	400	0	0	0	1	1	1	3	4	6	7	12	17
	500	0	0	0	1	1	1	2	3	5	6	10	14
	600	0	0	0	0	1	1	1	3	4	5	8	12
	700	0	0	0	0	1	1	1	2	3	4	7	10
	750	0	0	0	0	1	1	1	2	3	4	7	10
	800	0	0	0	0	1	1	1	2	3	4	6	9
	1,000	0	0	0	0	0	1	1	1	2	3	5	8
THHN, THWN, THWN-2	14	13	22	36	63	85	140	200	309	412	531	833	1,202
	12	9	16	26	46	62	102	146	225	301	387	608	877
	10	6	10	17	29	39	64	92	142	189	244	383	552
	8	3	6	9	16	22	37	53	82	109	140	221	318
	6	2	4	7	12	16	27	38	59	79	101	159	230
	4	1	2	4	7	10	16	23	36	48	62	98	141
	3	1	1	3	6	8	14	20	31	41	53	83	120
	2	1	1	3	5	7	11	17	26	34	44	70	100
	1	1	1	1	4	5	8	12	19	25	33	51	74

RIGID METAL CONDUIT—MAXIMUM NUMBER OF CONDUCTORS *(cont.)*

Type	Size	RMC Trade Size (in)											
		½	¾	1	1¼	1½	2	2½	3	3½	4	5	6
THHN,	1/0	1	1	1	3	4	7	10	16	21	27	43	63
THWN,	2/0	1	1	1	2	3	6	8	13	18	23	36	52
THWN-2	3/0	0	1	1	1	3	5	7	11	15	19	30	43
(cont.)	4/0	0	1	1	1	2	4	6	9	12	16	25	36
	250	0	0	1	1	1	3	5	7	10	13	20	29
	300	0	0	1	1	1	3	4	6	8	11	17	25
	350	0	0	1	1	1	2	3	5	7	10	15	22
	400	0	0	1	1	1	2	3	5	7	8	13	20
	500	0	0	0	1	1	1	2	4	5	7	11	16
	600	0	0	0	1	1	1	1	3	4	6	9	13
	700	0	0	0	1	1	1	1	3	4	5	8	11
	750	0	0	0	0	1	1	1	3	4	5	7	11
	800	0	0	0	0	1	1	1	2	3	4	7	10
	1,000	0	0	0	0	0	1	1	1	3	4	6	8
FEP, FEPB,	14	12	22	35	61	83	136	194	300	400	515	808	1,166
PFA, PFAH,	12	9	16	26	44	60	99	142	219	292	376	590	851
TFE	10	6	11	18	32	43	71	102	157	209	269	423	610
	8	3	6	10	18	25	41	58	90	120	154	242	350
	6	2	4	7	13	17	29	41	64	85	110	172	249
	4	1	3	5	9	12	20	29	44	59	77	120	174
	3	1	2	4	7	10	17	24	37	50	64	100	145
	2	1	1	3	6	8	14	20	31	41	53	83	120
PFA, PFAH, TFE	1	1	1	2	4	6	9	14	21	28	37	57	83
PFA, PFAH,	1/0	1	1	1	3	5	8	11	18	24	30	48	69
TFE, Z	2/0	1	1	1	3	4	6	9	14	19	25	40	57
	3/0	0	1	1	2	3	5	8	12	16	21	33	47
	4/0	0	1	1	1	2	4	6	10	13	17	27	39
Z	14	15	26	42	73	100	164	234	361	482	621	974	1,405
	12	10	18	30	52	71	116	166	256	342	440	691	997
	10	6	11	18	32	43	71	102	157	209	269	423	610
	8	4	7	11	20	27	45	64	99	132	170	267	386
	6	3	5	8	14	19	31	45	69	93	120	188	271
	4	1	3	5	9	13	22	31	48	64	82	129	186
	3	1	2	4	7	9	16	22	35	47	60	94	136
	2	1	1	3	6	8	13	19	29	39	50	78	113
	1	1	1	2	5	6	10	15	23	31	40	63	92
XHH,	14	9	15	25	44	59	98	140	216	288	370	581	839
XHHW,	12	7	12	19	33	45	75	107	165	221	284	446	644
XHHW-2,	10	5	9	14	25	34	56	80	123	164	212	332	480
ZW	8	3	5	8	14	19	31	44	68	91	118	185	267
	6	1	3	6	10	14	23	33	51	68	87	137	197
	4	1	2	4	7	10	16	24	37	49	63	99	143
	3	1	1	3	6	8	14	20	31	41	53	84	121
	2	1	1	3	5	7	12	17	26	35	45	70	101
	1	1	1	1	4	5	9	12	19	26	33	52	76
XHH,	1/0	1	1	1	3	4	7	10	16	22	28	44	64
XHHW,	2/0	0	1	1	2	3	6	9	13	18	23	37	53
XHHW-2	3/0	0	1	1	1	3	5	7	11	15	19	30	44
	4/0	0	1	1	1	2	4	6	9	12	16	25	36
	250	0	0	1	1	1	3	5	7	10	13	20	30
	300	0	0	1	1	1	3	4	6	8	11	18	25
	350	0	0	1	1	1	2	3	6	7	10	15	22
	400	0	0	1	1	1	2	3	5	7	9	14	20
	500	0	0	0	1	1	1	2	4	5	7	11	16
	600	0	0	0	1	1	1	1	3	4	6	9	13
	700	0	0	0	1	1	1	1	3	4	5	8	11
	750	0	0	0	0	1	1	1	3	4	5	7	11
	800	0	0	0	0	1	1	1	2	3	4	7	10

LIQUIDTIGHT FLEXIBLE METAL CONDUIT—MAXIMUM NUMBER OF CONDUCTORS

Type	Size	½	¾	1	1¼	1½	2	2½	3	3½	4
RHH, RHW,	14	4	7	12	21	27	44	66	102	133	173
RHW-2	12	3	6	10	17	22	36	55	84	110	144
	10	3	5	8	14	18	29	44	68	89	116
	8	1	2	4	7	9	15	23	36	46	61
	6	1	1	3	6	7	12	18	28	37	48
	4	1	1	2	4	6	9	14	22	29	38
	3	1	1	1	4	5	8	13	19	25	33
	2	1	1	1	3	4	7	11	17	22	29
	1	0	1	1	1	3	5	7	11	14	19
	1/0	0	1	1	1	2	4	6	10	13	16
	2/0	0	1	1	1	1	3	5	8	11	14
	3/0	0	0	1	1	1	3	4	7	9	12
	4/0	0	0	1	1	1	2	4	6	8	10
	250	0	0	0	1	1	1	3	4	6	8
	300	0	0	0	1	1	1	2	4	5	7
	350	0	0	0	1	1	1	2	3	5	6
	400	0	0	0	1	1	1	1	3	4	6
	500	0	0	0	1	1	1	1	3	4	5
	600	0	0	0	0	1	1	1	2	3	4
	700	0	0	0	0	0	1	1	1	3	3
	750	0	0	0	0	0	1	1	1	2	3
	800	0	0	0	0	0	1	1	1	2	3
	1,000	0	0	0	0	0	1	1	1	1	3
TW	14	9	15	25	44	57	93	140	215	280	365
	12	7	12	19	33	43	71	108	165	215	280
	10	5	9	14	25	32	53	80	123	160	209
	8	3	5	8	14	18	29	44	68	89	116
RHH*, RHW*,	14	6	10	16	29	38	62	93	143	186	243
RHW-2*, THHW,	12	5	8	13	23	30	50	75	115	149	195
THW, THW-2	10	3	6	10	18	23	39	58	89	117	152
	8	1	4	6	11	14	23	35	53	70	91
RHH*, RHW*,	6	1	3	5	8	11	18	27	41	53	70
RHW-2*, TW,	4	1	1	3	6	8	13	20	30	40	52
THW, THHW,	3	1	1	3	5	7	11	17	26	34	44
THW-2	2	1	1	2	4	6	9	14	22	29	38
	1	1	1	1	3	4	7	10	15	20	26
	1/0	0	1	1	2	3	6	8	13	17	23
	2/0	0	1	1	2	3	5	7	11	15	19
NOTE:	3/0	0	1	1	1	2	4	6	9	12	16
(*) Denotes	4/0	0	0	1	1	1	3	5	8	10	13
types minus	250	0	0	1	1	1	3	4	6	8	11
outer	300	0	0	1	1	1	2	3	5	7	9
covering	350	0	0	0	1	1	1	3	5	6	8
	400	0	0	0	1	1	1	3	4	6	7
	500	0	0	0	1	1	1	2	3	5	6
	600	0	0	0	1	1	1	1	3	4	5
	700	0	0	0	0	1	1	1	2	3	4
	750	0	0	0	0	1	1	1	2	3	4
	800	0	0	0	0	1	1	1	2	3	4
	1,000	0	0	0	0	0	1	1	1	2	3
THHN,	14	13	22	36	63	81	133	201	308	401	523
THWN,	12	9	16	26	46	59	97	146	225	292	381
THWN-2	10	6	10	16	29	37	61	92	141	184	240
	8	3	6	9	16	21	35	53	81	106	138
	6	2	4	7	12	15	25	38	59	76	100
	4	1	2	4	7	9	15	23	36	47	61
	3	1	1	3	6	8	13	20	30	40	52
	2	1	1	3	5	7	11	17	26	33	44
	1	1	1	1	4	5	8	12	19	25	32

LIQUIDTIGHT FLEXIBLE METAL CONDUIT – MAXIMUM NUMBER OF CONDUCTORS *(cont.)*

Type	Size	½	¾	1	1¼	1½	2	2½	3	3½	4
THHN, THWN, THWN-2 *(cont.)*	1/0	1	1	1	3	4	7	10	16	21	27
	2/0	0	1	1	2	3	6	8	13	17	23
	3/0	0	1	1	1	3	5	7	11	14	19
	4/0	0	1	1	1	2	4	6	9	12	15
	250	0	0	1	1	1	3	5	7	10	12
	300	0	0	1	1	1	3	4	6	8	11
	350	0	0	1	1	1	2	3	5	7	9
	400	0	0	0	1	1	1	3	5	6	8
	500	0	0	0	1	1	1	2	4	5	7
	600	0	0	0	0	1	1	1	3	4	6
	700	0	0	0	1	1	1	1	3	4	5
	750	0	0	0	0	1	1	1	3	3	5
	800	0	0	0	0	1	1	1	2	3	4
	1,000	0	0	0	0	0	1	1	1	3	3
FEP, FEPB, PFA, PFAH, TFE	14	12	21	35	61	79	129	195	299	389	507
	12	9	15	25	44	57	94	142	218	284	370
	10	6	11	18	32	41	68	102	156	203	266
	8	3	6	10	18	23	39	58	89	117	152
	6	2	4	7	13	17	27	41	64	83	108
	4	1	3	5	9	12	19	29	44	58	75
	3	1	2	4	7	10	16	24	37	48	63
	2	1	1	3	6	8	13	20	30	40	52
PFA, PFAH, TFE	1	1	1	2	4	5	9	14	21	28	36
PFA, PFAH, TFE, Z	1/0	1	1	1	3	4	7	11	18	23	30
	2/0	1	1	1	3	4	6	9	14	19	25
	3/0	0	1	1	2	3	5	8	12	16	20
	4/0	0	1	1	1	2	4	6	10	13	17
Z	14	20	26	42	73	95	156	235	360	469	611
	12	14	18	30	52	67	111	167	255	332	434
	10	8	11	18	32	41	68	102	156	203	266
	8	5	7	11	20	26	43	64	99	129	168
	6	4	5	8	14	18	30	45	69	90	118
	4	2	3	5	9	12	20	31	48	62	81
	3	2	2	4	7	9	15	23	35	45	59
	2	1	1	3	6	7	12	19	29	38	49
	1	1	1	2	5	6	10	15	23	30	40
XHH, XHHW, XHHW-2, ZW	14	9	15	25	44	57	93	140	215	280	365
	12	7	12	19	33	43	71	108	165	215	280
	10	5	9	14	25	32	53	80	123	160	209
	8	3	5	8	14	18	29	44	68	89	116
	6	1	3	6	10	13	22	33	50	66	86
	4	1	2	4	7	9	16	24	36	48	62
	3	1	1	3	6	8	13	20	31	40	52
	2	1	1	3	5	7	11	17	26	34	44
	1	1	1	1	4	5	8	12	19	25	33
XHH, XHHW, XHHW-2	1/0	1	1	1	3	4	7	10	16	21	28
	2/0	0	1	1	2	3	6	9	13	17	23
	3/0	0	1	1	1	3	5	7	11	14	19
	4/0	0	1	1	1	2	4	6	9	12	16
	250	0	0	1	1	1	3	5	7	10	13
	300	0	0	1	1	1	3	4	6	8	11
	350	0	0	1	1	1	2	3	5	7	10
	400	0	0	0	1	1	1	3	5	6	8
	500	0	0	0	1	1	1	2	4	5	7
	600	0	0	0	1	1	1	1	3	4	6
	700	0	0	0	1	1	1	1	3	4	5
	750	0	0	0	0	1	1	1	3	3	5
	800	0	0	0	0	1	1	1	2	3	4

NONMETALLIC TUBING—MAXIMUM NUMBER OF CONDUCTORS

Type	Size	ENT Trade Size (in)					
		½	¾	1	1¼	1½	2
RHH, RHW, RHW-2	14	3	6	10	19	26	43
	12	2	5	9	16	22	36
	10	1	4	7	13	17	29
	8	1	1	3	6	9	15
	6	1	1	3	5	7	12
	4	1	1	2	4	6	9
	3	1	1	1	3	5	8
	2	0	1	1	3	4	7
	1	0	1	1	1	3	5
	1/0	0	0	1	1	2	4
	2/0	0	0	1	1	1	3
	3/0	0	0	1	1	1	3
	4/0	0	0	1	1	1	2
	250	0	0	0	1	1	1
	300	0	0	0	1	1	1
	350	0	0	0	1	1	1
	400	0	0	0	1	1	1
	500	0	0	0	0	1	1
	600	0	0	0	0	1	1
	700	0	0	0	0	0	1
	750	0	0	0	0	0	1
	800	0	0	0	0	0	1
	1,000	0	0	0	0	0	1
TW	14	7	13	22	40	55	92
	12	5	10	17	31	42	71
	10	4	7	13	23	32	52
	8	1	4	7	13	17	29
RHH*, RHW*, RHW-2*, THHW, THW, THW-2	14	4	8	15	27	37	61
	12	3	7	12	21	29	49
	10	3	5	9	17	23	38
	8	1	3	5	10	14	23
RHH*, RHW*, RHW-2*, TW, THW, THHW, THW-2	6	1	2	4	7	10	17
	4	1	1	3	5	8	13
	3	1	1	2	5	7	11
	2	1	1	2	4	6	9
	1	0	1	1	3	4	6
	1/0	0	1	1	2	3	5
	2/0	0	1	1	1	3	5
NOTE: (*) Denotes types minus outer covering	3/0	0	0	1	1	2	4
	4/0	0	0	1	1	1	3
	250	0	0	1	1	1	2
	300	0	0	0	1	1	2
	350	0	0	0	1	1	1
	400	0	0	0	1	1	1
	500	0	0	0	1	1	1
	600	0	0	0	0	1	1
	700	0	0	0	0	1	1
	750	0	0	0	0	1	1
	800	0	0	0	0	1	1
	1,000	0	0	0	0	0	1
THHN, THWN, THWN-2	14	10	18	32	58	80	132
	12	7	13	23	42	58	96
	10	4	8	15	26	36	60
	8	2	5	8	15	21	35
	6	1	3	6	11	15	25
	4	1	1	4	7	9	15
	3	1	1	3	5	8	13
	2	1	1	2	5	6	11
	1	1	1	1	3	5	8

NONMETALLIC TUBING—MAXIMUM NUMBER OF CONDUCTORS (cont.)

Type	Size	ENT Trade Size (in) ½	¾	1	1¼	1½	2
THHN,	1/0	0	1	1	3	4	7
THWN,	2/0	0	1	1	2	3	5
THWN-2	3/0	0	1	1	1	3	4
(cont.)	4/0	0	0	1	1	2	4
	250	0	0	1	1	1	3
	300	0	0	1	1	1	2
	350	0	0	0	1	1	2
	400	0	0	0	1	1	1
	500	0	0	0	1	1	1
	600	0	0	0	1	1	1
	700	0	0	0	0	1	1
	750	0	0	0	0	1	1
	800	0	0	0	0	1	1
	1,000	0	0	0	0	0	1
FEP, FEPB,	14	10	18	31	56	77	128
PFA, PFAH,	12	7	13	23	41	56	93
TFE	10	5	9	16	29	40	67
	8	3	5	9	17	23	38
	6	1	4	6	12	16	27
	4	1	2	4	8	11	19
	3	1	1	4	7	9	16
	2	1	1	3	5	8	13
PFA, PFAH, TFE	1	1	1	1	4	5	9
PFA, PFAH,	1/0	0	1	1	3	4	7
TFE, Z	2/0	0	1	1	2	4	6
	3/0	0	1	1	1	3	5
	4/0	0	1	1	1	2	4
Z	14	12	22	38	68	93	154
	12	8	15	27	48	66	109
	10	5	9	16	29	40	67
	8	3	6	10	18	25	42
	6	1	4	7	13	18	30
	4	1	3	5	9	12	20
	3	1	1	3	6	9	15
	2	1	1	3	5	7	12
	1	1	1	2	4	6	10
XHH,	14	7	13	22	40	55	92
XHHW,	12	5	10	17	31	42	71
XHHW-2,	10	4	7	13	23	32	52
ZW	8	1	4	7	13	17	29
	6	1	3	5	9	13	21
	4	1	1	4	7	9	15
	3	1	1	3	6	8	13
	2	1	1	2	5	6	11
XHH,	1	1	1	1	3	5	8
XHHW,	1/0	0	1	1	3	4	7
XHHW-2	2/0	0	1	1	2	3	6
	3/0	0	1	1	1	3	5
	4/0	0	0	1	1	2	4
	250	0	0	1	1	1	3
	300	0	0	1	1	1	3
	350	0	0	1	1	1	2
	400	0	0	0	1	1	1
	500	0	0	0	1	1	1
	600	0	0	0	1	1	1
	700	0	0	0	0	1	1
	750	0	0	0	0	1	1
	800	0	0	0	0	1	1

ELECTRICAL METALLIC TUBING–MAXIMUM NUMBER OF CONDUCTORS

Type	Size	EMT Trade Size (in) ½	¾	1	1¼	1½	2	2½	3	3½	4
RHH, RHW,	14	4	7	11	20	27	46	80	120	157	201
RHW-2	12	3	6	9	17	23	38	66	100	131	167
	10	2	5	8	13	18	30	53	81	105	135
	8	1	2	4	7	9	16	28	42	55	70
	6	1	1	3	5	8	13	22	34	44	56
	4	1	1	2	4	6	10	17	26	34	44
	3	1	1	1	4	5	9	15	23	30	38
	2	1	1	1	3	4	7	13	20	26	33
	1	0	1	1	1	3	5	9	13	17	22
	1/0	0	1	1	1	2	4	7	11	15	19
	2/0	0	1	1	1	2	4	6	10	13	17
	3/0	0	0	1	1	1	3	5	8	11	14
	4/0	0	0	0	1	1	3	5	7	9	12
	250	0	0	0	1	1	1	3	5	7	9
	300	0	0	0	1	1	1	3	5	6	8
	350	0	0	0	1	1	1	3	4	6	7
	400	0	0	0	1	1	1	2	4	5	7
	500	0	0	0	0	1	1	2	3	4	6
	600	0	0	0	0	1	1	1	3	4	5
	700	0	0	0	0	0	1	1	2	3	4
	750	0	0	0	0	0	1	1	2	3	4
	800	0	0	0	0	0	1	1	2	3	4
	1,000	0	0	0	0	0	1	1	1	2	3
TW	14	8	15	25	43	58	96	168	254	332	424
	12	6	11	19	33	45	74	129	195	255	326
	10	5	8	14	24	33	55	96	145	190	243
	8	2	5	8	13	18	30	53	81	105	135
RHH*, RHW*,	14	6	10	16	28	39	64	112	169	221	282
RHW-2*, THHW,	12	4	8	13	23	31	51	90	136	177	227
THW, THW-2,	10	3	6	10	18	24	40	70	106	138	177
	8	1	4	6	10	14	24	42	63	83	106
RHH*, RHW*,	6	1	3	4	8	11	18	32	48	63	81
RHW-2*, TW,	4	1	1	3	6	8	13	24	36	47	60
THW, THHW,	3	1	1	3	5	7	12	20	31	40	52
THW-2	2	1	1	2	4	6	10	17	26	34	44
	1	1	1	1	3	4	7	12	18	24	31
	1/0	0	1	1	2	3	6	10	16	20	26
	2/0	0	1	1	1	3	5	9	13	17	22
	3/0	0	1	1	1	2	4	7	11	15	19
	4/0	0	0	1	1	1	3	6	9	12	16
	250	0	0	1	1	1	3	5	7	10	13
	300	0	0	1	1	1	2	4	6	8	11
	350	0	0	0	1	1	1	4	6	7	10
	400	0	0	0	1	1	1	3	5	7	9
NOTE:	500	0	0	0	1	1	1	3	4	6	7
(*) Denotes	600	0	0	0	1	1	1	2	3	4	6
types minus	700	0	0	0	0	1	1	1	3	4	5
outer	750	0	0	0	0	1	1	1	3	4	5
covering	800	0	0	0	0	1	1	1	3	3	5
	1,000	0	0	0	0	0	1	1	3	3	4
THHN,	14	12	22	35	61	84	138	241	364	476	608
THWN,	12	9	16	26	45	61	101	176	266	347	443
THWN-2	10	5	10	16	28	38	63	111	167	219	279
	8	3	6	9	16	22	36	64	96	126	161
	6	2	4	7	12	16	26	46	69	91	116
	4	1	2	4	7	10	16	28	43	56	71
	3	1	1	3	6	8	13	24	36	47	60
	2	1	1	3	5	7	11	20	30	40	51
	1	1	1	1	4	5	8	15	22	29	37

ELECTRICAL METALLIC TUBING – MAXIMUM NUMBER OF CONDUCTORS (cont.)

Type	Size	EMT Trade Size (in)									
		½	¾	1	1¼	1½	2	2½	3	3½	4
THHN, THWN, THWN-2 (cont.)	1/0	1	1	1	3	4	7	12	19	25	32
	2/0	0	1	1	2	3	6	10	16	20	26
	3/0	0	1	1	1	3	5	8	13	17	22
	4/0	0	1	1	1	2	4	7	11	14	18
	250	0	0	1	1	1	3	6	9	11	15
	300	0	0	1	1	1	3	5	7	10	13
	350	0	0	1	1	1	2	4	6	9	11
	400	0	0	0	1	1	1	4	6	8	10
	500	0	0	0	1	1	1	3	5	6	8
	600	0	0	0	1	1	1	2	4	5	7
	700	0	0	0	1	1	1	2	3	4	6
	750	0	0	0	0	1	1	1	3	4	5
	800	0	0	0	0	1	1	1	3	4	5
	1,000	0	0	0	0	1	1	1	2	3	4
FEP, FEPB, PFA, PFAH, TFE	14	12	21	34	60	81	134	234	354	462	590
	12	9	15	25	43	59	98	171	258	337	430
	10	6	11	18	31	42	70	122	185	241	309
	8	3	6	10	18	24	40	70	106	138	177
	6	2	4	7	12	17	28	50	75	98	126
	4	1	3	5	9	12	20	35	53	69	88
	3	1	2	4	7	10	16	29	44	57	73
	2	1	1	3	6	8	13	24	36	47	60
PFA, PFAH, TFE	1	1	1	2	4	6	9	16	25	33	42
PFA, PFAH, TFE, Z	1/0	1	1	1	3	5	8	14	21	27	35
	2/0	0	1	1	3	4	6	11	17	22	29
	3/0	0	1	1	2	3	5	9	14	18	24
	4/0	0	1	1	1	2	4	8	11	15	19
Z	14	14	25	41	72	98	161	282	426	556	711
	12	10	18	29	51	69	114	200	302	394	504
	10	6	11	18	31	42	70	122	185	241	309
	8	4	7	11	20	27	44	77	117	153	195
	6	3	5	8	14	19	31	54	82	107	137
	4	1	3	5	9	13	21	37	56	74	94
	3	1	2	4	7	9	15	27	41	54	69
	2	1	1	3	6	8	13	22	34	45	57
	1	1	1	2	4	6	10	18	28	36	46
XHH, XHHW, XHHW-2, ZW	14	8	15	25	43	58	96	168	254	332	424
	12	6	11	19	33	45	74	129	195	255	326
	10	5	8	14	24	33	55	96	145	190	243
	8	2	5	8	13	18	30	53	81	105	135
	6	1	3	6	10	14	22	39	60	78	100
	4	1	2	4	7	10	16	28	43	56	72
	3	1	1	3	6	8	14	24	36	48	61
	2	1	1	3	5	7	11	20	31	40	51
XHH, XHHW, XHHW-2	1	1	1	1	4	5	8	15	23	30	38
	1/0	1	1	1	3	4	7	13	19	25	32
	2/0	0	1	1	2	3	6	10	16	21	27
	3/0	0	1	1	1	3	5	9	13	17	22
	4/0	0	1	1	1	2	4	7	11	14	18
	250	0	0	1	1	1	3	6	9	12	15
	300	0	0	1	1	1	3	5	8	10	13
	350	0	0	1	1	1	2	4	7	9	11
	400	0	0	0	1	1	1	4	6	8	10
	500	0	0	0	1	1	1	3	5	6	8
	600	0	0	0	1	1	1	2	4	5	6
	700	0	0	0	0	1	1	1	2	3	4
	750	0	0	0	0	1	1	1	3	4	5
	800	0	0	0	0	1	1	1	3	4	5

FLEXIBLE METAL CONDUIT – MAXIMUM NUMBER OF CONDUCTORS

Type	Size	FMC Trade Size (in)									
		½	¾	1	1¼	1½	2	2½	3	3½	4
RHH, RHW,	14	4	7	11	17	25	44	67	96	131	171
RHW-2	12	3	6	9	14	21	37	55	80	109	142
	10	3	5	7	11	17	30	45	64	88	115
	8	1	2	4	6	9	15	23	34	46	60
	6	1	1	3	5	7	12	19	27	37	48
	4	1	1	2	4	5	10	14	21	29	37
	3	1	1	1	3	5	8	13	18	25	33
	2	1	1	1	3	4	7	11	16	22	28
	1	0	1	1	1	2	5	7	10	14	19
	1/0	0	1	1	1	2	4	6	9	12	16
	2/0	0	1	1	1	1	3	5	8	11	14
	3/0	0	0	1	1	1	3	5	7	9	12
	4/0	0	0	1	1	1	2	4	6	8	10
	250	0	0	0	1	1	1	3	4	6	8
	300	0	0	0	1	1	1	2	4	5	7
	350	0	0	0	1	1	1	2	3	5	6
	400	0	0	0	0	1	1	1	3	4	6
	500	0	0	0	0	1	1	1	3	4	5
	600	0	0	0	0	1	1	1	2	3	4
	700	0	0	0	0	0	1	1	1	3	3
	750	0	0	0	0	0	1	1	1	3	3
	800	0	0	0	0	0	1	1	1	2	3
	1,000	0	0	0	0	0	1	1	1	1	3
TW	14	9	15	23	36	53	94	141	203	277	361
	12	7	11	18	28	41	72	108	156	212	277
	10	5	8	13	21	30	54	81	116	158	207
	8	3	5	7	11	17	30	45	64	88	115
RHH*, RHW*,	14	6	10	15	24	35	62	94	135	184	240
RHW-2*, THHW,	12	5	8	12	19	28	50	75	108	148	193
THW, THW-2	10	4	6	10	15	22	39	59	85	115	151
	8	1	4	6	9	13	23	35	51	69	90
RHH*, RHW*,	6	1	3	4	7	10	18	27	39	53	69
RHW-2*, TW,	4	1	1	3	5	7	13	20	29	39	51
THW, THHW,	3	1	1	3	4	6	11	17	25	34	44
THW-2	2	1	1	2	4	5	10	14	21	29	37
	1	1	1	1	2	4	7	10	15	20	26
	1/0	0	1	1	1	3	6	9	12	17	22
	2/0	0	1	1	1	3	5	7	10	14	19
NOTE:	3/0	0	1	1	1	2	4	6	9	12	16
(*) Denotes	4/0	0	0	1	1	1	3	5	7	10	13
types minus	250	0	0	1	1	1	3	4	6	8	11
outer	300	0	0	1	1	1	2	3	5	7	9
covering	350	0	0	0	1	1	1	3	4	6	8
	400	0	0	0	1	1	1	3	4	6	7
	500	0	0	0	1	1	1	2	3	5	6
	600	0	0	0	0	1	1	1	3	4	5
	700	0	0	0	0	1	1	1	2	3	4
	750	0	0	0	0	1	1	1	2	3	4
	800	0	0	0	0	1	1	1	1	3	4
	1,000	0	0	0	0	0	1	1	1	2	3
THHN,	14	13	22	33	52	76	134	202	291	396	518
THWN,	12	9	16	24	38	56	98	147	212	289	378
THWN-2	10	6	10	15	24	35	62	93	134	182	238
	8	3	6	9	14	20	35	53	77	105	137
	6	2	4	6	10	14	25	38	55	76	99
	4	1	2	4	6	9	16	24	34	46	61
	3	1	1	3	5	7	13	20	29	39	51
	2	1	1	3	4	6	11	17	24	33	43
	1	1	1	1	3	4	8	12	18	24	32

FLEXIBLE METAL CONDUIT – MAXIMUM NUMBER OF CONDUCTORS (cont.)

Type	AWG/kcmil	½	¾	1	1¼	1½	2	2½	3	3½	4
THHN, THWN, THWN-2 (cont.)	1/0	1	1	1	2	4	7	10	15	20	27
	2/0	0	1	1	1	3	6	9	12	17	22
	3/0	0	1	1	1	2	5	7	10	14	18
	4/0	0	1	1	1	1	4	6	8	12	15
	250	0	0	1	1	1	3	5	7	9	12
	300	0	0	1	1	1	3	4	6	8	11
	350	0	0	1	1	1	2	3	5	7	9
	400	0	0	0	1	1	1	3	5	6	8
	500	0	0	0	1	1	1	2	4	5	7
	600	0	0	0	0	1	1	1	3	4	5
	700	0	0	0	0	1	1	1	3	4	5
	750	0	0	0	0	1	1	1	2	3	4
	800	0	0	0	0	1	1	1	2	3	4
	1,000	0	0	0	0	0	1	1	1	3	3
FEP, FEPB, PFA, PFAH, TFE	14	12	21	32	51	74	130	196	282	385	502
	12	9	15	24	37	54	95	143	206	281	367
	10	6	11	17	26	39	68	103	148	201	263
	8	4	6	10	15	22	39	59	85	115	151
	6	2	4	7	11	16	28	42	60	82	107
	4	1	3	5	7	11	19	29	42	57	75
	3	1	2	4	6	9	16	24	35	48	62
	2	1	1	3	5	7	13	20	29	39	51
PFA, PFAH, TFE	1	1	1	2	3	5	9	14	20	27	36
PFA, PFAH, TFE, Z	1/0	1	1	1	3	4	8	11	17	23	30
	2/0	1	1	1	2	3	6	9	14	19	24
	3/0	0	1	1	1	3	5	8	11	15	20
	4/0	0	1	1	1	2	4	6	9	13	16
Z	14	15	25	39	61	89	157	236	340	463	605
	12	11	18	28	43	63	111	168	241	329	429
	10	6	11	17	26	39	68	103	148	201	263
	8	4	7	11	17	24	43	65	93	127	166
	6	3	5	7	12	17	30	45	65	89	117
	4	1	3	5	8	12	21	31	45	61	80
	3	1	2	4	6	8	15	23	33	45	58
	2	1	1	3	5	7	12	19	27	37	49
	1	1	1	2	4	6	10	15	22	30	39
XHH, XHHW, XHHW-2, ZW	14	9	15	23	36	53	94	141	203	277	361
	12	7	11	18	28	41	72	108	156	212	277
	10	5	8	13	21	30	54	81	116	158	207
	8	3	5	7	11	17	30	45	64	88	115
	6	1	3	5	8	12	22	33	48	65	85
	4	1	2	4	6	9	16	24	34	47	61
	3	1	1	3	5	7	13	20	29	40	52
	2	1	1	3	4	6	11	17	24	33	44
XHH, XHHW, XHHW-2	1	1	1	1	3	5	8	13	18	25	32
	1/0	1	1	1	2	4	7	10	15	21	27
	2/0	0	1	1	2	3	6	9	13	17	23
	3/0	0	1	1	1	3	5	7	10	14	19
	4/0	0	1	1	1	2	4	6	9	12	15
	250	0	0	1	1	1	3	5	7	10	13
	300	0	0	1	1	1	3	4	6	8	11
	350	0	0	1	1	1	2	4	5	7	9
	400	0	0	0	1	1	1	3	5	6	8
	500	0	0	0	1	1	1	3	4	5	7
	600	0	0	0	0	1	1	1	3	4	5
	700	0	0	0	0	1	1	1	3	4	5
	750	0	0	0	0	1	1	1	2	3	4
	800	0	0	0	0	1	1	1	2	3	4

RIGID PVC SCHEDULE 40 CONDUIT – MAXIMUM NUMBER OF CONDUCTORS

Type	Size	PVC 40 Trade Size (in)											
		½	¾	1	1¼	1½	2	2½	3	3½	4	5	6
RHH, RHW, RHW-2	14	4	7	11	20	27	45	64	99	133	171	269	390
	12	3	5	9	16	22	37	53	82	110	142	224	323
	10	2	4	7	13	18	30	43	66	89	115	181	261
	8	1	2	4	7	9	15	22	35	46	60	94	137
	6	1	1	3	5	7	12	18	28	37	48	76	109
	4	1	1	2	4	6	10	14	22	29	37	59	85
	3	1	1	1	4	5	8	12	19	25	33	52	75
	2	1	1	1	3	4	7	10	16	22	28	45	65
	1	0	1	1	1	3	5	7	11	14	19	29	43
	1/0	0	1	1	1	2	4	6	9	13	16	26	37
	2/0	0	0	1	1	1	3	5	8	11	14	22	32
	3/0	0	0	1	1	1	3	4	7	9	12	19	28
	4/0	0	0	1	1	1	2	4	6	8	10	16	24
	250	0	0	0	1	1	1	3	4	6	8	12	18
	300	0	0	0	1	1	1	2	4	5	7	11	16
	350	0	0	0	1	1	1	2	3	5	6	10	14
	400	0	0	0	1	1	1	1	3	4	6	9	13
	500	0	0	0	0	1	1	1	3	4	5	8	11
	600	0	0	0	0	1	1	1	2	3	4	6	9
	700	0	0	0	0	0	1	1	1	3	3	6	8
	750	0	0	0	0	0	1	1	1	2	3	5	8
	800	0	0	0	0	0	1	1	1	2	3	5	7
	1,000	0	0	0	0	0	1	1	1	1	2	4	6
TW	14	8	14	24	42	57	94	135	209	280	361	568	822
	12	6	11	18	32	44	72	103	160	215	277	436	631
	10	4	8	13	24	32	54	77	119	160	206	325	470
	8	2	4	7	13	18	30	43	66	89	115	181	261
RHH*, RHW*, RHW-2*, THHW, THW, THW-2	14	5	9	16	28	38	63	90	139	186	240	378	546
	12	4	8	12	22	30	50	72	112	150	193	304	439
	10	3	6	10	17	24	39	56	87	117	150	237	343
	8	1	3	6	10	14	23	33	52	70	90	142	205
RHH*, RHW*, RHW-2*, TW, THW, THHW, THW-2	6	1	2	4	8	11	18	26	40	53	69	109	157
	4	1	1	3	6	8	13	19	30	40	51	81	117
	3	1	1	3	5	7	11	16	25	34	44	69	100
	2	1	1	2	4	6	10	14	22	29	37	59	85
	1	0	1	1	3	4	7	10	15	20	26	41	60
	1/0	0	1	1	2	3	6	8	13	17	22	35	51
	2/0	0	1	1	1	3	5	7	11	15	19	30	43
NOTE:	3/0	0	1	1	1	2	4	6	9	12	16	25	36
(*) Denotes	4/0	0	0	1	1	1	3	5	8	10	13	21	30
types minus	250	0	0	1	1	1	3	4	6	8	11	17	25
outer	300	0	0	1	1	1	2	3	5	7	9	15	21
covering	350	0	0	0	1	1	1	3	5	6	8	13	19
	400	0	0	0	1	1	1	3	4	6	7	12	17
	500	0	0	0	1	1	1	2	3	5	6	10	14
	600	0	0	0	0	1	1	1	3	4	5	8	11
	700	0	0	0	0	1	1	1	2	3	4	7	10
	750	0	0	0	0	1	1	1	2	3	4	6	10
	800	0	0	0	0	1	1	1	2	3	4	6	9
	1,000	0	0	0	0	0	1	1	1	2	3	5	7
THHN, THWN, THWN-2	14	11	21	34	60	82	135	193	299	401	517	815	1,178
	12	8	15	25	43	59	99	141	218	293	377	594	859
	10	5	9	15	27	37	62	89	137	184	238	374	541
	8	3	5	9	16	21	36	51	79	106	137	216	312
	6	1	4	6	11	15	26	37	57	77	99	156	225
	4	1	2	4	7	9	16	22	35	47	61	96	138
	3	1	1	3	6	8	13	19	30	40	51	81	117
	2	1	1	3	5	7	11	16	25	33	43	68	98
	1	1	1	1	3	5	8	12	18	25	32	50	73

RIGID PVC SCHEDULE 40 CONDUIT – MAXIMUM NUMBER OF CONDUCTORS (cont.)

Type	Size	\frac{1}{2}	\frac{3}{4}	1	1\frac{1}{4}	1\frac{1}{2}	2	2\frac{1}{2}	3	3\frac{1}{2}	4	5	6
						PVC 40 Trade Size (in)							
THHN, THWN, THWN-2 (cont.)	1/0	1	1	1	3	4	7	10	15	21	27	42	61
	2/0	0	1	1	2	3	6	8	13	17	22	35	51
	3/0	0	1	1	1	3	5	7	11	14	18	29	42
	4/0	0	1	1	1	2	4	6	9	12	15	24	35
	250	0	0	1	1	1	3	4	7	10	12	20	28
	300	0	0	1	1	1	3	4	6	8	11	17	24
	350	0	0	1	1	1	2	3	5	7	9	15	21
	400	0	0	0	1	1	1	3	5	6	8	13	19
	500	0	0	0	1	1	1	2	4	5	7	11	16
	600	0	0	0	1	1	1	1	3	4	5	9	13
	700	0	0	0	0	1	1	1	3	4	5	8	11
	750	0	0	0	0	1	1	1	2	3	4	7	11
	800	0	0	0	0	1	1	1	2	3	4	7	10
	1,000	0	0	0	0	0	1	1	1	3	3	6	8
FEP, FEPB, PFA, PFAH, TFE	14	11	20	33	58	79	131	188	290	389	502	790	1,142
	12	8	15	24	42	58	96	137	212	284	366	577	834
	10	6	10	17	30	41	69	98	152	204	263	414	598
	8	3	6	10	17	24	39	56	87	117	150	237	343
	6	2	4	7	12	17	28	40	62	83	107	169	244
	4	1	3	5	8	12	19	28	43	58	75	118	170
	3	1	2	4	7	10	16	23	36	48	62	98	142
	2	1	1	3	6	8	13	19	30	40	51	81	117
PFA, PFAH, TFE	1	1	1	2	4	5	9	13	20	28	36	56	81
PFA, PFAH, TFE, Z	1/0	1	1	1	3	4	8	11	17	23	30	47	68
	2/0	0	1	1	3	4	6	9	14	19	24	39	56
	3/0	0	1	1	2	3	5	7	12	16	20	32	46
	4/0	0	1	1	1	2	4	6	9	13	16	26	38
Z	14	13	24	40	70	95	158	226	350	469	605	952	1,376
	12	9	17	28	49	68	112	160	248	333	429	675	976
	10	6	10	17	30	41	69	98	152	204	263	414	598
	8	3	6	11	19	26	43	62	96	129	166	261	378
	6	2	4	7	13	18	30	43	67	90	116	184	265
	4	1	3	5	9	12	21	30	46	62	80	126	183
	3	1	2	4	6	9	15	22	34	45	58	92	133
	2	1	1	3	5	7	12	18	28	38	49	77	111
	1	1	1	2	4	6	10	14	23	30	39	62	90
XHH, XHHW, XHHW-2, ZW	14	8	14	24	42	57	94	135	209	280	361	568	822
	12	6	11	18	32	44	72	103	160	215	277	436	631
	10	4	8	13	24	32	54	77	119	160	206	325	470
	8	2	4	7	13	18	30	43	66	89	115	181	261
	6	1	3	5	10	13	22	32	49	66	85	134	193
	4	1	2	4	7	9	16	23	35	48	61	97	140
	3	1	1	3	6	8	13	19	30	40	52	82	118
	2	1	1	3	5	7	11	16	25	34	44	69	99
XHH, XHHW, XHHW-2	1	1	1	1	3	5	8	12	19	25	32	51	74
	1/0	1	1	1	3	4	7	10	16	21	27	43	62
	2/0	0	1	1	2	3	6	8	13	17	23	36	52
	3/0	0	1	1	1	3	5	7	11	14	19	30	43
	4/0	0	1	1	1	2	4	6	9	12	15	24	35
	250	0	0	1	1	1	3	5	7	10	13	20	29
	300	0	0	1	1	1	3	4	6	8	11	17	25
	350	0	0	1	1	1	2	3	5	7	9	15	22
	400	0	0	0	1	1	1	3	5	6	8	13	19
	500	0	0	0	1	1	1	2	4	5	7	11	16
	600	0	0	0	1	1	1	1	3	4	5	9	13
	700	0	0	0	0	1	1	1	3	4	5	8	11
	750	0	0	0	0	1	1	1	2	3	4	7	11
	800	0	0	0	0	1	1	1	2	3	4	6	10

RIGID PVC SCHEDULE 80 CONDUIT – MAXIMUM NUMBER OF CONDUCTORS

Type	Size	\½	¾	1	1¼	1½	2	2½	3	3½	4	5	6
						PVC 80 Trade Size (in)							
RHH, RHW, RHW-2	14	3	5	9	17	23	39	56	88	118	153	243	349
	12	2	4	7	14	19	32	46	73	98	127	202	290
	10	1	3	6	11	15	26	37	59	79	103	163	234
	8	1	1	3	6	8	13	19	31	41	54	85	122
	6	1	1	2	4	6	11	16	24	33	43	68	98
	4	1	1	1	3	5	8	12	19	26	33	53	77
	3	0	1	1	3	4	7	11	17	23	29	47	67
	2	0	1	1	3	4	6	9	14	20	25	41	58
	1	0	1	1	1	2	4	6	9	13	17	27	38
	1/0	0	0	1	1	1	3	5	8	11	15	23	33
	2/0	0	0	1	1	1	3	4	7	10	13	20	29
	3/0	0	0	1	1	1	3	4	6	8	11	17	25
	4/0	0	0	0	1	1	2	3	5	7	9	15	21
	250	0	0	0	1	1	1	2	4	5	7	11	16
	300	0	0	0	1	1	1	2	3	5	6	10	14
	350	0	0	0	0	1	1	1	3	4	5	9	13
	400	0	0	0	0	1	1	1	3	4	5	8	12
	500	0	0	0	0	0	1	1	2	3	4	7	10
	600	0	0	0	0	0	0	1	1	3	3	6	8
	700	0	0	0	0	0	0	1	1	2	3	5	7
	750	0	0	0	0	0	0	1	1	2	3	5	7
	800	0	0	0	0	0	0	1	1	2	3	4	7
	1,000	0	0	0	0	0	1	1	1	1	2	4	5
TW	14	6	11	20	35	49	82	118	185	250	324	514	736
	12	5	9	15	27	38	63	91	142	192	248	394	565
	10	3	6	11	20	28	47	67	106	143	185	294	421
	8	1	3	6	11	15	26	37	59	79	103	163	234
RHH*, RHW*, RHW-2*, THHW THW, THW-2	14	4	8	13	23	32	55	79	123	166	215	341	490
	12	3	6	10	19	26	44	63	99	133	173	274	394
	10	2	5	8	15	20	34	49	77	104	135	214	307
	8	1	3	5	9	12	20	29	46	62	81	128	184
RHH*, RHW*, RHW-2*, TW, THW, THHW, THW-2	6	1	1	3	7	9	16	22	35	48	62	98	141
	4	1	1	3	5	7	12	17	26	35	46	73	105
	3	1	1	2	4	6	10	14	22	30	39	63	90
	2	1	1	1	3	5	8	12	19	26	33	53	77
	1	0	1	1	2	3	6	8	13	18	23	37	54
	1/0	0	1	1	1	3	5	7	11	15	20	32	46
	2/0	0	1	1	1	2	4	6	10	13	17	27	39
	3/0	0	0	1	1	1	3	5	8	11	14	23	33
	4/0	0	0	1	1	1	3	4	7	9	12	19	27
	250	0	0	0	1	1	2	3	5	7	9	15	22
	300	0	0	0	1	1	1	3	5	6	8	13	19
	350	0	0	0	1	1	1	2	4	6	7	12	17
	400	0	0	0	1	1	1	2	4	5	7	10	15
	500	0	0	0	0	1	1	1	3	4	5	9	13
	600	0	0	0	0	1	1	1	2	3	4	7	10
	700	0	0	0	0	1	1	1	2	3	4	6	9
	750	0	0	0	0	0	1	1	1	3	4	6	8
	800	0	0	0	0	0	1	1	1	3	3	6	8
	1,000	0	0	0	0	0	1	1	1	2	3	5	7
THHN, THWN, THWN-2	14	9	17	28	51	70	118	170	265	358	464	736	1,055
	12	6	12	20	37	51	86	124	193	261	338	537	770
	10	4	7	13	23	32	54	78	122	164	213	338	485
	8	2	4	7	13	18	31	45	70	95	123	195	279
	6	1	3	5	9	13	22	32	51	68	89	141	202
	4	1	1	3	6	8	14	20	31	42	54	86	124
	3	1	1	3	5	7	12	17	26	35	46	73	105
	2	1	1	2	4	6	10	14	22	30	39	61	88
	1	0	1	1	3	4	7	10	16	22	29	45	65

NOTE: (*) Denotes types minus outer covering

RIGID PVC SCHEDULE 80 CONDUIT – MAXIMUM NUMBER OF CONDUCTORS (cont.)

Type	Size	PVC 80 Trade Size (in)											
		½	¾	1	1¼	1½	2	2½	3	3½	4	5	6
THHN,	1/0	0	1	1	2	3	6	9	14	18	24	38	55
THWN,	2/0	0	1	1	1	3	5	7	11	15	20	32	46
THWN-2	3/0	0	1	1	1	2	4	6	9	13	17	26	38
(cont.)	4/0	0	0	1	1	1	3	5	8	10	14	22	31
	250	0	0	1	1	1	3	4	6	8	11	18	25
	300	0	0	0	1	1	2	3	5	7	9	15	22
	350	0	0	0	1	1	1	3	5	6	8	13	19
	400	0	0	0	1	1	1	3	4	6	7	12	17
	500	0	0	0	1	1	1	2	3	5	6	10	14
	600	0	0	0	0	1	1	1	3	4	5	8	12
	700	0	0	0	0	1	1	1	2	3	4	7	10
	750	0	0	0	0	1	1	1	2	3	4	7	9
	800	0	0	0	0	1	1	1	2	3	4	6	9
	1,000	0	0	0	0	0	1	1	1	2	3	5	7
FEP, FEPB,	14	8	16	27	49	68	115	164	257	347	450	714	1,024
PFA, PFAH,	12	6	12	20	36	50	84	120	188	253	328	521	747
TFE	10	4	8	14	26	36	60	86	135	182	235	374	536
	8	2	5	8	15	20	34	49	77	104	135	214	307
	6	1	3	6	10	14	24	35	55	74	96	152	218
	4	1	2	4	7	10	17	24	38	52	67	106	153
	3	1	1	3	6	8	14	20	32	43	56	89	127
	2	1	1	3	5	7	12	17	26	35	46	73	105
PFA, PFAH, TFE	1	1	1	1	3	5	8	11	18	25	32	51	73
PFA, PFAH,	1/0	0	1	1	3	4	7	10	15	20	27	42	61
TFE, Z	2/0	0	1	1	2	3	5	8	12	17	22	35	50
	3/0	0	1	1	1	2	4	6	10	14	18	29	41
	4/0	0	0	1	1	1	4	5	8	11	15	24	34
Z	14	10	19	33	59	82	138	198	310	418	542	860	1,233
	12	7	14	23	42	58	98	141	220	297	385	610	875
	10	4	8	14	26	36	60	86	135	182	235	374	536
	8	3	5	9	16	22	38	54	85	115	149	236	339
	6	2	4	6	11	16	26	38	60	81	104	166	238
	4	1	2	4	8	11	18	26	41	55	72	114	164
	3	1	2	3	5	8	13	19	30	40	52	83	119
	2	1	1	2	5	6	11	16	25	33	43	69	99
	1	0	1	2	4	5	9	13	20	27	35	56	80
XHH,	14	6	11	20	35	49	82	118	185	250	324	514	736
XHHW,	12	5	9	15	27	38	63	91	142	192	248	394	565
XHHW-2,	10	3	6	11	20	28	47	67	106	143	185	294	421
ZW	8	1	3	6	11	15	26	37	59	79	103	163	234
	6	1	2	4	8	11	19	28	43	59	76	121	173
	4	1	1	3	6	8	14	20	31	42	55	87	125
	3	1	1	3	5	7	12	17	26	36	47	74	106
	2	1	1	2	4	6	10	14	22	30	39	62	89
XHH, XHHW,	1	0	1	1	3	4	7	10	16	22	29	46	66
XHHW-2	1/0	0	1	1	2	3	6	9	14	19	24	39	56
	2/0	0	1	1	1	3	5	7	11	16	20	32	46
	3/0	0	1	1	1	2	4	6	9	13	17	27	38
	4/0	0	0	1	1	1	3	5	8	11	14	22	32
	250	0	0	1	1	1	3	4	6	9	11	18	26
	300	0	0	1	1	1	2	3	5	7	10	15	22
	350	0	0	0	1	1	1	3	5	6	8	14	20
	400	0	0	0	1	1	1	3	4	6	7	12	17
	500	0	0	0	1	1	1	2	3	5	6	10	14
	600	0	0	0	0	1	1	1	3	4	5	8	11
	700	0	0	0	0	1	1	1	2	3	4	7	10
	750	0	0	0	0	1	1	1	1	3	4	6	9
	800	0	0	0	0	1	1	1	1	3	4	5	9

COPPER BUS-BAR DATA
Sizes, Weights, and Resistances

Thickness (in)	Width (in)	weights per ft at 0.3213 lb per cu in	Area (sq in)	ohms per ft at 8.341 per sq mil. ft	Capacity (amperes)
1/16	1/2	.1205	.0313	.00026691	30
1/16	3/4	.1807	.0469	.00017790	50
1/16	1	.2410	.0625	.00013344	60
1/16	1-1/2	.3615	.0938	.00008897	90
1/8	1/2	.2410	.0625	.00013344	75
1/8	3/4	.3615	.0938	.00008897	90
1/8	1	.4820	.125	.00006672	125
1/8	1-1/2	.7230	.1875	.00004448	200
1/8	2	.9640	.25	.00003336	250
1/4	3/4	.7230	.1875	.00004448	185
1/4	1	.9640	.25	.00003336	250
1/4	1-1/4	1.205	.3125	.00002669	315
1/4	1-1/2	1.446	.375	.00002224	375
1/4	1-3/4	1.687	.4375	.00001906	435
1/4	2	1.928	.5	.00001668	500
1/4	2-1/4	2.169	.5625	.00001482	565
1/4	2-1/2	2.410	.625	.00001334	630
1/2	3/4	1.446	.375	.00002224	370
1/2	1	1.928	.500	.00001668	500
1/2	1-1/4	2.410	.625	.00001334	625
1/2	1-1/2	2.892	.750	.00001112	750
1/2	1-3/4	3.374	.875	.00000953	875
1/2	2	3.856	1.	.00000834	1,000
1/2	2-1/4	4.338	1.125	.00000741	1,185
1/2	2-1/2	4.820	1.25	.00000667	1,250
1/2	2-3/4	5.304	1.375	.00000606	1,375
1/2	3	5.784	1.500	.00000556	1,500
1/2	3-1/4	6.266	1.625	.00000513	1,625
1/2	3-1/2	6.748	1.750	.00000475	1,750
1/2	3-3/4	7.23	1.875	.00000444	1,875
1/2	4	7.712	2.000	.00000417	2,000
3/4	1	2.892	.750	.00001112	750
3/4	1-1/2	4.338	1.125	.00000741	1,125
3/4	2	5.784	1.500	.00000556	1,500
3/4	2-1/2	7.23	1.875	.00000444	1,875
3/4	3	8.676	2.250	.00000370	2,250
3/4	3-1/2	10.122	2.625	.00000317	2,650
3/4	4	11.568	3.000	.00000278	3,000

Carrying capacity is figured at 1,000 amperes per square inch.

COMPARATIVE WEIGHTS OF COPPER AND ALUMINUM CONDUCTORS lb per 1,000 ft

Wire Size	Bare Solid			Bare Stranded		
	Copper	Alum.	Diff.	Copper	Alum.	Diff.
18	4.92	1.49	3.43	5.02	1.53	3.49
16	7.82	2.38	5.44	7.97	2.43	5.54
14	12.43	3.78	8.65	12.68	3.86	8.82
12	19.77	6.01	13.76	20.16	6.13	14.03
10	31.43	9.55	21.88	32.06	9.75	22.31
8	50.0	15.2	34.8	51.0	15.5	35.5
6	79.5	24.2	55.3	81.0	24.6	56.4
4	126.4	38.4	88.0	128.9	39.2	89.7
2	200.9	60.8	140.1	204.9	62.3	142.6
1	253.3	77.0	176.3	258.4	78.6	179.8
1/0	319.5	97.2	222.3	325.8	99.1	226.7
2/0	402.8	122.6	280.2	410.9	124.9	286.0
3/0	507.9	154.6	353.3	518.1	157.5	360.6
4/0	640.5	194.9	445.6	653.3	198.6	454.7
250	–	–	–	771.9	234.7	537.2
300	–	–	–	926.3	281.6	644.7
350	–	–	–	1,081	328.6	752
400	–	–	–	1,235	375.5	859
450	–	–	–	1,389	422.4	967
500	–	–	–	1,544	469.4	1,075
600	–	–	–	1,853	563	1,290
700	–	–	–	2,161	657	1,504
750	–	–	–	2,316	704	1,612
800	–	–	–	2,470	751	1,719
900	–	–	–	2,779	845	1,934
1,000	–	–	–	3,088	939	2,149
1,250	–	–	–	3,859	1,173	2,686
1,500	–	–	–	4,631	1,410	3,221
1,750	–	–	–	5,403	1,645	3,758
2,000	–	–	–	6,175	1,880	4,295

COPPER WIRE SPECIFICATIONS

Wire Size (AWG)	Area (Circular Mils)	Diameter (Mils, 1,000th in)	Diameter (mm)	Weight (lb per 1,000 ft)
40	9.9	3.1	.080	.0200
38	15.7	4	.101	.0476
36	25	5	.127	.0757
34	39.8	6.3	.160	.120
32	63.2	8	.202	.191
30	101	10	.255	.304
28	160	12.6	.321	.484
26	254	15.9	.405	.769
24	404	20.1	.511	1.22
22	642	25.3	.644	1.94
20	1,020	32	.812	3.09
18	1,620	40	1.024	4.92
16	2,580	51	1.291	7.82
14	4,110	64	1.628	12.4
12	6,530	81	2.053	19.8
10	10,400	102	2.588	31.4
8	16,500	128	3.264	50
6	26,300	162	4.115	79.5
4	41,700	204	5.189	126
3	52,600	229	5.827	159
2	66,400	258	6.544	201
1	83,700	289	7.348	253
1/0	106,000	325	8.255	319
2/0	133,000	365	9.271	403
3/0	168,000	410	10.414	508
4/0	212,000	460	11.684	641

COPPER WIRE RESISTANCE

Wire Size (AWG)	77°F (ft per ohm)	149°F (ft per ohm)	77°F (ohms per 1,000 ft)	149°F (ohms per 1,000 ft)
40	.93	.81	1,070	1,230
38	1.5	1.3	673	776
36	2.4	2.0	423	488
34	3.8	3.3	266	307
32	6.0	5.2	167	193
30	9.5	8.3	105	121
28	15.1	13.1	66.2	76.4
26	24.0	20.8	41.6	48.0
24	38.2	33.1	26.2	30.2
22	60.6	52.6	16.5	19.0
20	96.2	84.0	10.4	11.9
18	153.6	133.2	6.51	7.51
16	244.5	211.4	4.09	4.73
14	387.6	336.7	2.58	2.97
12	617.3	534.8	1.62	1.87
10	980.4	847.5	1.02	1.18
8	1,560	1,353	.641	.739
6	2,481	2,151	.403	.465
4	3,953	3,425	.253	.292
3	4,975	4,310	.201	.232
2	6,289	5,435	.159	.184
1	7,936	6,849	.126	.146
1/0	10,000	8,621	.100	.116
2/0	12,658	10,870	.079	.092
3/0	15,873	13,699	.063	.073
4/0	20,000	17,544	.050	.057

CONDUCTOR DIMENSIONS AND RESISTANCES

Wire Size	Area (Cir. Mils)	Stranding Wire Qty.	Stranding Dia. (in)	Overall Dimensions Dia. (in)	Overall Dimensions Area (in²)	DC Resistance, ohm/K ft @ 75°C Copper Uncoated	DC Resistance, ohm/K ft @ 75°C Copper Coated	DC Resistance, ohm/K ft @ 75°C Alum.
18	1,620	1	–	0.040	0.001	7.77	8.08	12.8
18	1,620	7	0.015	0.046	0.002	7.95	8.45	13.1
16	2,580	1	–	0.051	0.002	4.89	5.08	8.05
16	2,580	7	0.019	0.058	0.003	4.99	5.29	8.21
14	4,110	1	–	0.064	0.003	3.07	3.19	5.06
14	4,110	7	0.024	0.073	0.004	3.14	3.26	5.17
12	6,530	1	–	0.081	0.005	1.93	2.01	3.18
12	6,530	7	0.030	0.092	0.006	1.98	2.05	3.25
10	10,380	1	–	0.102	0.008	1.21	1.26	2.00
10	10,380	7	0.038	0.116	0.011	1.24	1.29	2.04
8	16,510	1	–	0.128	0.013	0.764	0.786	1.26
8	16,510	7	0.049	0.146	0.017	0.778	0.809	1.28

CONDUCTOR DIMENSIONS AND RESISTANCES (cont.)

Wire Size	Area (Cir. Mils)	Stranding		Overall Dimensions		DC Resistance, ohm/K ft @ 75°C		
		Wire Qty.	Dia. (in)	Dia. (in)	Area (in²)	Copper Uncoated	Copper Coated	Alum.
6	26,240	7	0.061	0.184	0.027	0.491	0.510	0.808
4	41,740	7	0.077	0.232	0.042	0.308	0.321	0.508
3	52,620	7	0.087	0.260	0.053	0.245	0.254	0.403
2	66,360	7	0.097	0.292	0.067	0.194	0.201	0.319
1	83,690	19	0.066	0.332	0.087	0.154	0.160	0.253
1/0	105,600	19	0.074	0.372	0.109	0.122	0.127	0.201
2/0	133,100	19	0.084	0.418	0.137	0.0967	0.101	0.159
3/0	167,800	19	0.094	0.470	0.173	0.0766	0.0797	0.126
4/0	211,600	19	0.106	0.528	0.219	0.0608	0.0626	0.100
250	250,000	37	0.082	0.575	0.260	0.0515	0.0535	0.0847
300	300,000	37	0.090	0.630	0.312	0.0429	0.0446	0.0707
350	350,000	37	0.097	0.681	0.364	0.0367	0.0382	0.0605

CONDUCTOR DIMENSIONS AND RESISTANCES *(cont.)*

| Wire Size | Area (Cir. Mils) | Stranding | | Overall Dimensions | | DC Resistance ohm/K ft @ 75°C | | |
| | | Wire Qty. | Dia. (in) | Dia. (in) | Area (in²) | Copper | | Alum. |
						Uncoated	Coated	
400	400,000	37	0.104	0.728	0.416	0.0321	0.0331	0.0529
500	500,000	37	0.116	0.813	0.519	0.0258	0.0265	0.0424
600	600,000	61	0.099	0.893	0.626	0.0214	0.0223	0.0353
700	700,000	61	0.107	0.964	0.730	0.0184	0.0189	0.0303
750	750,000	61	0.111	0.998	0.782	0.0171	0.0176	0.0282
800	800,000	61	0.114	1.030	0.834	0.0161	0.0166	0.0265
900	900,000	61	0.122	1.094	0.940	0.0143	0.0147	0.0235
1,000	1,000,000	61	0.128	1.152	1.042	0.0129	0.0132	0.0212
1,250	1,250,000	91	0.117	1.289	1.305	0.0103	0.0106	0.0169
1,500	1,500,000	91	0.128	1.412	1.566	0.00858	0.00883	0.0141
1,750	1,750,000	127	0.117	1.526	1.829	0.00735	0.00756	0.0121
2,000	2,000,000	127	0.126	1.632	2.092	0.00643	0.00662	0.0106

AC RESISTANCE, REACTANCE, & IMPEDANCE FOR 1,000 VOLT CABLES MEASURED IN OHMS TO NEUTRAL PER 1,000 FEET, 3 PHASE, 60 Hz, 75°C BASED ON THREE SINGLE CONDUCTORS IN PVC OR STEEL CONDUIT

Wire Size	AC Resistance for Uncoated Copper Wires		AC Resistance for Aluminum Wires		Reactance for All Wires		Impedance at 0.85 PF for Uncoated Copper Wires		Impedance at 0.85 PF for Aluminum Wires	
	PVC	Steel	PVC	Steel	PVC	Steel	PVC	Steel	PVC	Steel
14	3.1	3.1	–	–	0.058	0.073	2.7	2.7	–	–
12	2.0	2.0	3.2	3.2	0.054	0.068	1.7	1.7	2.8	2.8
10	1.2	1.2	2.0	2.0	0.050	0.063	1.1	1.1	1.8	1.8
8	0.78	0.78	1.3	1.3	0.052	0.065	0.69	0.70	1.1	1.1
6	0.49	0.49	0.81	0.81	0.051	0.064	0.44	0.45	0.71	0.72
4	0.31	0.31	0.51	0.51	0.048	0.060	0.29	0.30	0.46	0.46
3	0.25	0.25	0.40	0.40	0.047	0.059	0.23	0.24	0.37	0.37

AC RESISTANCE, REACTANCE, & IMPEDANCE FOR 1,000 VOLT CABLES MEASURED IN OHMS TO NEUTRAL PER 1,000 FEET, 3 PHASE, 60 Hz, 75°C BASED ON THREE SINGLE CONDUCTORS IN PVC OR STEEL CONDUIT *(cont.)*

Wire Size	AC Resistance for Uncoated Copper Wires		AC Resistance for Aluminum Wires		Reactance for All Wires		Impedance at 0.85 PF for Uncoated Copper Wires		Impedance at 0.85 PF for Aluminum Wires	
	PVC	Steel	PVC	Steel	PVC	Steel	PVC	Steel	PVC	Steel
2	0.19	0.20	0.32	0.32	0.045	0.057	0.19	0.20	0.30	0.30
1	0.15	0.16	0.25	0.25	0.046	0.057	0.16	0.16	0.24	0.25
1/0	0.12	0.12	0.20	0.20	0.044	0.055	0.13	0.13	0.19	0.20
2/0	0.10	0.10	0.16	0.16	0.043	0.054	0.11	0.11	0.16	0.16
3/0	0.077	0.079	0.13	0.13	0.042	0.052	0.088	0.094	0.13	0.14
4/0	0.062	0.063	0.10	0.10	0.041	0.051	0.074	0.080	0.11	0.11
250	0.052	0.054	0.085	0.086	0.041	0.052	0.066	0.073	0.094	0.10

AC RESISTANCE, REACTANCE, & IMPEDANCE FOR 1,000 VOLT CABLES MEASURED IN OHMS TO NEUTRAL PER 1,000 FEET, 3 PHASE, 60 Hz, 75°C BASED ON THREE SINGLE CONDUCTORS IN PVC OR STEEL CONDUIT (cont.)

Wire Size	AC Resistance for Uncoated Copper Wires		AC Resistance for Aluminum Wires		Reactance for All Wires		Impedance at 0.85 PF for Uncoated Copper Wires		Impedance at 0.85 PF for Aluminum Wires	
	PVC	Steel	PVC	Steel	PVC	Steel	PVC	Steel	PVC	Steel
300	0.044	0.045	0.071	0.072	0.041	0.051	0.059	0.065	0.082	0.088
350	0.038	0.039	0.061	0.063	0.040	0.050	0.053	0.060	0.073	0.080
400	0.033	0.035	0.054	0.055	0.040	0.049	0.049	0.056	0.066	0.073
500	0.027	0.029	0.043	0.045	0.039	0.048	0.043	0.050	0.057	0.064
600	0.023	0.025	0.036	0.038	0.039	0.048	0.040	0.047	0.051	0.058
750	0.019	0.021	0.029	0.031	0.038	0.048	0.036	0.043	0.045	0.052
1,000	0.015	0.018	0.023	0.025	0.037	0.046	0.032	0.040	0.039	0.046

AMPACITIES OF COPPER CONDUCTORS (1)

Ampacities of Single Insulated Conductors Rated 0–2,000 Volts in Free Air

Wire Size	Ambient Temperature 30°C (86°F)			Ambient Temperature 40°C (104°F)			Bare Conductors with Max Temp 80°C (176°F)
	60°C (140°F)	75°C (167°F)	90°C (194°F)	150°C (302°F)	200°C (392°F)	250°C (482°F)	
	Types	Types	Types	Type	Types	Types	
AWG kcmil	TW UF	RHW, THHW, THW, THWN, XHHW, ZW	TBS, SA, SIS, FEP, MI, FEPB, ZW-2, RHH, RHW-2, THHN, THHW, THW-2, XHH, USE-2, THWN-2, XHHW, XHHW-2	Z	FEP, FEPB, PFA, SA	PFAH, TFE Nickel or nickel-coated copper	
14*	25	30	35	46	54	59	—
12*	30	35	40	60	68	78	—
10*	40	50	55	80	90	107	—
8	60	70	80	106	124	142	98
6	80	95	105	155	165	205	124
4	105	125	140	190	220	278	155
3	120	145	165	214	252	327	—
2	140	170	190	255	293	381	209
1	165	195	220	293	344	440	—
1/0	195	230	260	339	399	532	282
2/0	225	265	300	390	467	591	329
3/0	260	310	350	451	546	708	382
4/0	300	360	405	529	629	830	444
250	340	405	455	—	—	—	494

Temperature Correction Factors

Ambient Temp.°C	For Other Than 30°C			For Other Than 40°C			Ambient Temp.°C
	Multiply the Ampacities above by the Factors Below						
21–25	1.08	1.05	1.04	.95	.97	.98	41–50
26–30	1.00	1.00	1.00	.90	.94	.95	51–60
31–35	.91	.94	.96	.85	.90	.93	61–70
36–40	.82	.88	.91	.80	.87	.90	71–80
41–45	.71	.82	.87	.74	.83	.87	81–90
46–50	.58	.75	.82	.67	.79	.85	91–100
51–55	.41	.67	.76	.52	.71	.79	101–120
56–60	—	.58	.71	.30	.61	.72	121–140
61–70	—	.33	.58	—	.50	.65	141–160
71–80	—	—	.41	—	.35	.58	161–180
					—	.49	181–200
					—	.35	201–225

*Unless specifically permitted by the NEC®, overcurrent protection for copper conductors shall not exceed 15 amps for no. 14 AWG, 20 amps for no. 12 AWG, and 30 amps for no. 10 AWG.

AMPACITIES OF COPPER CONDUCTORS (1) *(cont.)*

Ampacities of Single Insulated Conductors Rated 0–2,000 Volts in Free Air

Wire Size	Ambient Temperature 30°C (86°F)			Ambient Temperature 40°C (104°F)			Bare Conductors with Max Temp 80°C (176°F)
	60°C (140°F)	75°C (167°F)	90°C (194°F)	150°C (302°F)	200°C (392°F)	250°C (482°F)	
	Types TW, UF	Types RHW, THHW, THW, THWN, XHHW, ZW	Types TBS, SA, SIS, FEP, MI, FEPB, ZW-2, RHH, RHW-2, THHN, THHW, THW-2, XHH, USE-2, THWN-2, XHHW, XHHW-2.	Type Z	Types FEP, FEPB, PFA, SA	Types PFAH, TFE Nickel or nickel-coated copper	
kcmil							
300	375	445	505	—	—	—	556
350	420	505	570	—	—	—	—
400	455	545	615	—	—	—	—
500	515	620	700	—	—	—	773
600	575	690	780	—	—	—	—
700	630	755	855	—	—	—	—
750	655	785	885	—	—	—	1,000
800	680	815	920	—	—	—	—
900	730	870	985	—	—	—	—
1,000	780	935	1,055	—	—	—	1,193
1,250	890	1,065	1,200	—	—	—	—
1,500	980	1,175	1,325	—	—	—	—
1,750	1,070	1,280	1,445	—	—	—	—
2,000	1,155	1,385	1,560	—	—	—	—

Temperature Correction Factors			
Ambient Temp.°C	For Other Than 30°C, Multiply the Ampacities above by the Factors Below		
21–25	1.08	1.05	1.04
26–30	1.00	1.00	1.00
31–35	.91	.94	.96
36–40	.82	.88	.91
41–45	.71	.82	.87
46–50	.58	.75	.82
51–55	.41	.67	.76
56–60	—	.58	.71
61–70	—	.33	.58
71–80	—	—	.41

AMPACITIES OF COPPER CONDUCTORS (3)

Ampacities of Not More Than 3 Insulated Conductors Rated 0–2,000 Volts

Wire Size AWG kcmil	In Cable, Raceway, or Earth, Ambient Temperature 30°C (86°F)			In Cable or Raceway, Ambient Temperature 40°C (104°F)			Wire Size AWG
	60°C (140°F)	75°C (167°F)	90°C (194°F)	150°C (302°F)	200°C (392°F)	250°C (482°F)	
	Types TW UF	Types RHW, THHW, THW, THWN, XHHW, USE, ZW	Types TBS, SA, SIS, FEP, MI FEPB, ZW-2, RHH, RHW-2, THHN, THHW, THW-2, XHH, USE-2, THWN-2, XHHW, XHHW-2	Type Z	Types FEP, FEPB, PFA, SA	Types PFAH, TFE Nickel or nickel-coated copper	
14*	20	20	25	34	36	39	14
12*	25	25	30	43	45	54	12
10*	30	35	40	55	60	73	10
8	40	50	55	76	83	93	8
6	55	65	75	96	110	117	6
4	70	85	95	120	125	148	4
3	85	100	110	143	152	166	3
2	95	115	130	160	171	191	2
1	110	130	150	186	197	215	1
1/0	125	150	170	215	229	244	1/0
2/0	145	175	195	251	260	273	2/0
3/0	165	200	225	288	297	308	3/0
4/0	195	230	260	332	346	361	4/0
250	215	255	290	—	—	—	

Temperature Correction Factors

Ambient Temp.°C	For Other Than 30°C			For Other Than 40°C			Ambient Temp.°C
	Multiply the Ampacities above by the Factors Below						
21–25	1.08	1.05	1.04	.95	.97	.98	41–50
26–30	1.00	1.00	1.00	.90	.94	.95	51–60
31–35	.91	.94	.96	.85	.90	.93	61–70
36–40	.82	.88	.91	.80	.87	.90	71–80
41–45	.71	.82	.87	.74	.83	.87	81–90
46–50	.58	.75	.82	.67	.79	.85	91–100
51–55	.41	.67	.76	.52	.71	.79	101–120
56–60	—	.58	.71	.30	.61	.72	121–140
61–70	—	.33	.58	—	.50	.65	141–160
71–80	—	—	.41	—	.35	.58	161–180
				—	—	.49	181–200
				—	—	.35	201–225

*Unless specifically permitted by the NEC®, overcurrent protection for copper conductors shall not exceed 15 amps for no. 14 AWG, 20 amps for no. 12 AWG, and 30 amps for no. 10 AWG.

AMPACITIES OF COPPER CONDUCTORS (3) *(cont.)*

Ampacities of Not More Than 3 Insulated Conductors Rated 0–2,000 Volts

Wire Size	In Cable, Raceway, or Earth, Ambient Temperature 30°C (86°F)			In Cable or Raceway, Ambient Temperature 40°C (104°F)			Wire Size
	60°C (140°F)	75°C (167°F)	90°C (194°F)	150°C (302°F)	200°C (392°F)	250°C (482°F)	
kcmil	Types TW, UF	Types RHW, THHW, THW, THWN, XHHW, USE, ZW	Types TBS, SA, SIS, FEP, MI FEPB, ZW-2, RHH, RHW-2, THHN, THHW, THW-2, XHH, USE-2, THWN-2, XHHW, XHHW-2	Type Z	Types FEP, FEPB, PFA, SA	Types PFAH, TFE Nickel or nickel-coated copper	
300	240	285	320				
350	260	310	350				
400	280	335	380				
500	320	380	430				
600	355	420	475				
700	385	460	520				
750	400	475	535				
800	410	490	555				
900	435	520	585				
1,000	455	545	615				
1,250	495	590	665				
1,500	520	625	705				
1,750	545	650	735				
2,000	560	665	750				

Temperature Correction Factors

Ambient Temp.°C	For Other Than 30°C, Multiply the Ampacities above by the Factors Below		
21–25	1.08	1.05	1.04
26–30	1.00	1.00	1.00
31–35	.91	.94	.96
36–40	.82	.88	.91
41–45	.71	.82	.87
46–50	.58	.75	.82
51–55	.41	.67	.76
56–60	—	.58	.71
61–70	—	.33	.58
71–80	—	—	.41

AMPACITIES OF ALUMINUM OR COPPER-CLAD ALUMINUM CONDUCTORS (1)

Ampacities of Single Insulated Conductors Rated 0–2,000 Volts in Free Air

Wire Size	Ambient Temperature 30°C (86°F)			Ambient Temp. 40°C (104°F)	Wire Size
	60°C (140°F)	75°C (167°F)	90°C (194°F)	150°C (302°F)	
AWG kcmil	Types	Types	Types	Type	**AWG**
	TW, UF	THWN, THHW, XHHW, RHW, THW	TBS, XHH, RHH, RHW-2, XHHW, USE-2, THHN, SA, THHW, THW-2, XHHW-2, ZW-2, THWN-2, SIS	Z	
12*	25	30	35	47	12
10*	35	40	40	63	10
8	45	55	60	83	8
6	60	75	80	112	6
4	80	100	110	148	4
3	95	115	130	170	3
2	110	135	150	198	2
1	130	155	175	228	1
1/0	150	180	205	263	1/0
2/0	175	210	235	305	2/0
3/0	200	240	275	351	3/0
4/0	235	280	315	411	4/0
250	265	315	355		

Temperature Correction Factors

Ambient Temp.°C	For Other Than 30°C			40°C	Ambient Temp.°C
	Multiply the Ampacities above by the Factors Below				
21–25	1.08	1.05	1.04	.95	41–50
26–30	1.00	1.00	1.00	.90	51–60
31–35	.91	.94	.96	.85	61–70
36–40	.82	.88	.91	.80	71–80
41–45	.71	.82	.87	.74	81–90
46–50	.58	.75	.82	.67	91–100
51–55	.41	.67	.76	.52	101–120
56–60	—	.58	.71	.30	121–140
61–70	—	.33	.58		
71–80	—	—	.41		

*Unless specifically permitted by the NEC®, overcurrent protection for aluminum and copper-clad aluminum conductors shall not exceed 15 amps for no. 12 AWG and 25 amps for no. 10 AWG.

AMPACITIES OF ALUMINUM OR
COPPER-CLAD ALUMINUM CONDUCTORS (1) *(cont.)*

Ampacities of Single Insulated Conductors Rated 0–2,000 Volts in Free Air

	Ambient Temperature 30°C (86°F)			Ambient Temp. 40°C (104°F)	
Wire Size	60°C (140°F)	75°C (167°F)	90°C (194°F)	150°C (302°F)	Wire Size
	Types	Types	Types	Type	
kcmil	TW, UF	THWN, THHW, XHHW, RHW, THW	TBS, XHH RHH, RHW-2, XHHW, USE-2, THHN, SA, THHW, THW-2, XHHW-2, ZW-2, THWN-2, SIS	Z	
300	290	350	395		
350	330	395	445		
400	355	425	480		
500	405	485	545		
600	455	540	615		
700	500	595	675		
750	515	620	700		
800	535	645	725		
900	580	700	785		
1,000	625	750	845		
1,250	710	855	960		
1,500	795	950	1,075		
1,750	875	1,050	1,185		
2,000	960	1,150	1,335		

Temperature Correction Factors					
Ambient Temp.°C	For Other Than 30°C, Multiply the Ampacities above by the Factors Below				
21–25	1.08	1.05	1.04		
26–30	1.00	1.00	1.00		
31–35	.91	.94	.96		
36–40	.82	.88	.91		
41–45	.71	.82	.87		
46–50	.58	.75	.82		
51–55	.41	.67	.76		
56–60	—	.58	.71		
61–70	—	.33	.58		
71–80	—	—	.41		

AMPACITIES OF ALUMINUM AND COPPER-CLAD ALUMINUM CONDUCTORS (3)

Ampacities of Not More Than 3 Insulated Conductors Rated 0–2,000 Volts

Wire Size	In Cable, Raceway, or Earth, Ambient Temperature 30°C (86°F)			In Cable or Raceway, Ambient Temp. 40°C (104°F)	Wire Size
	60°C (140°F)	75°C (167°F)	90°C (194°F)	150°C (302°F)	
	Types	Types	Types	Type	
AWG kcmil	TW, UF	THWN, THHW, XHHW, USE, THW, RHW	TBS, SA, SIS, THHN, THHW, THW-2, THWN-2, RHH, RHW-2, USE-2, XHH, XHHW, XHHW-2, ZW-2	Z	AWG
12*	20	20	25	30	12
10*	25	30	35	44	10
8	30	40	45	57	8
6	40	50	60	75	6
4	55	65	75	94	4
3	65	75	85	109	3
2	75	90	100	124	2
1	85	100	115	145	1
1/0	100	120	135	169	1/0
2/0	115	135	150	198	2/0
3/0	130	155	175	227	3/0
4/0	150	180	205	260	4/0
250	170	205	230		

Temperature Correction Factors

Ambient Temp.°C	For Other Than 30°C			40°C	Ambient Temp.°C
	Multiply the Ampacities above by the Factors Below				
21–25	1.08	1.05	1.04	.95	41–50
26–30	1.00	1.00	1.00	.90	51–60
31–35	.91	.94	.96	.85	61–70
36–40	.82	.88	.91	.80	71–80
41–45	.71	.82	.87	.74	81–90
46–50	.58	.75	.82	.67	91–100
51–55	.41	.67	.76	.52	101–120
56–60	—	.58	.71	.30	121–140
61–70	—	.33	.58		
71–80	—	—	.41		

*Unless specifically permitted by the NEC®, overcurrent protection for aluminum and copper-clad aluminum conductors shall not exceed 15 amps for no. 12 AWG and 25 amps for no. 10 AWG.

AMPACITIES OF ALUMINUM AND
COPPER-CLAD ALUMINUM CONDUCTORS (3) *(cont.)*

Ampacities of Not More Than 3 Insulated Conductors Rated 0–2,000 Volts

Wire Size	In Cable, Raceway, or Earth, Ambient Temperature 30°C (86°F)			In Cable or Raceway Ambient Temp. 40°C (104°F)	Wire Size
	60°C (140°F)	75°C (167°F)	90°C (194°F)	150°C (302°F)	
	Types	Types	Types	Type	
kcmil	TW, UF	THWN, THHW, XHHW, USE, THW, RHW	TBS, SA, SIS, THHN, THHW, THW-2, THWN-2, RHH, RHW-2, USE-2, XHH, XHHW, XHHW-2, ZW-2	Z	
300	190	230	255		
350	210	250	280		
400	225	270	305		
500	260	310	350		
600	285	340	385		
700	310	375	420		
750	320	385	435		
800	330	395	450		
900	355	425	480		
1,000	375	445	500		
1,250	405	485	545		
1,500	435	520	585		
1,750	455	545	615		
2,000	470	560	630		

Temperature Correction Factors					
Ambient Temp.°C	For Other Than 30°C, Multiply the Ampacities above by the Factors Below				
21–25	1.08	1.05	1.04		
26–30	1.00	1.00	1.00		
31–35	.91	.94	.96		
36–40	.82	.88	.91		
41–45	.71	.82	.87		
46–50	.58	.75	.82		
51–55	.41	.67	.76		
56–60	—	.58	.71		
61–70	—	.33	.58		
71–80	—	—	.41		

AMPACITY RATINGS FOR FLEXIBLE CORDS AND POWER CABLES AT 30°C AMBIENT TEMPERATURE

Wire Size (AWG)	Types C, E, S, SJ, SJO, SJOW, SJOO, SO, SOW, SOO, SP-1, SP-2, SP-3, SRD, SJT, SJTW, SPE-1, SPE-2, SPE-3, SPT-1, SPT-1W, SPT-2, SPT-2W, SPT-3, ST, SRDE, SRDT, STO, STOW, STOO, STOOW, SVE, SVEO, SVT, SVTO, SVTOO		Types HPD, HPN, HSJ, HSJO, HSJOC
	A*	B*	
18	7	10	10
17	—	12	13
16	10	13	15
15	—	—	17
14	15	18	20
12	20	25	30
10	25	30	35
8	35	40	—
6	45	55	—
4	60	70	—
2	80	95	—

A*—For Three (3) current-carrying conductors

B*—For Two (2) current-carrying conductors

AMPACITY ADJUSTMENTS FOR 4 OR MORE CONDUCTORS IN A CABLE OR RACEWAY

Number of Current-Carrying Conductors	Percent of Values in Ampacity Charts/Adjust for Ambient Temperature (if Necessary)
4 to 6	80%
7 to 9	70%
10 to 20	50%
21 to 30	45%
31 to 40	40%
41 and more	35%

Note: For use with ampacity charts on pages 3–42 through 3–50.

AMPACITY RATINGS FOR SINGLE-PHASE SERVICE OR FEEDER CONDUCTORS IN NORMAL DWELLING UNITS

Service or Feeder Rating in Amps	Copper	Aluminum or Copper-Clad Aluminum
100	4 AWG	2 AWG
110	3	1
125	2	1/0
150	1	2/0
175	1/0	3/0
200	2/0	4/0
225	3/0	250 kcmil
250	4/0	300
300	250 kcmil	350
350	350	500
400	400	600

BRANCH CIRCUIT REQUIREMENTS

Branch Circuit Rating in Amps	Minimum Circuit Conductor Size	Minimum Tap Conductor Size	Overcurrent Protection Rating in Amps	Basic Electrical Lampholder Type	Outlet Receptacle Rating in Amps
15	14 AWG	14 AWG	15	Any	15 max
20	12 AWG	14 AWG	20	Any	15 or 20
30	10 AWG	14 AWG	30	Heavy Duty	30
40	8 AWG	12 AWG	40	Heavy Duty	40 or 50
50	6 AWG	12 AWG	50	Heavy Duty	50

BRANCH CIRCUIT PERMISSIBLE LOADS

The load shall never exceed the branch-circuit amp rating. An individual branch circuit shall be permitted to supply any load for which it is rated. A branch circuit supplying two or more outlets or receptacles shall supply only the loads specified according to its size as follows:

(1) 15- or 20-amp branch circuits shall be permitted to supply lighting, other equipment, or a combination of both, except small-appliance branch circuits, laundry branch circuits, and bathroom branch circuits required in dwelling units and shall supply only those receptacle outlets.

   (A) The rating of any one cord-and-plug-connected utilization equipment not fastened in place shall not exceed 80 percent of the branch-circuit amp rating.

   (B) The total rating of utilization equipment fastened in place, other than luminaires, shall not exceed 50 percent of the branch-circuit amp rating.

(2) 30-amp branch circuits shall be permitted to supply utilization equipment in any occupancy. A rating of any one cord-and-plug-connected utilization equipment shall not exceed 80 percent of the branch-circuit amp rating.

(3) 40- or 50-amp branch circuits shall be permitted to supply cooking appliances that are fastened in place in any occupancy. Larger than 50 amp shall supply only nonlighting outlet loads.

BRANCH CIRCUIT FIXTURE WIRE TAP SIZES

20-amp circuits — 18 AWG, up to a 50 ft run

20-amp circuits — 16 AWG, up to a 100 ft run

20-amp circuits — 14 AWG and larger

30-amp circuits — 14 AWG and larger

40-amp circuits — 12 AWG and larger

50-amp circuits — 12 AWG and larger

MAXIMUM CORD-AND-PLUG-CONNECTED LOAD TO RECEPTACLE

Circuit Rating (amps)	Recep. Rating (amps)	Maximum Load (amps)
15 or 20	15	12
20	20	16
30	30	24

RECEPTACLE RATINGS FOR VARIOUS SIZE CIRCUITS

Circuit Rating (amps)	Receptacle Rating (amps)
15	Not over 15
20	15 or 20
30	30
40	40 or 50
50	50

DEMAND FACTORS FOR HOUSEHOLD ELECTRIC CLOTHES DRYERS

Dryers	Demand Factor
1 to 4	100%
5	85%
6	75%
7	65%
8	60%
9	55%
10	50%

DEMAND FACTORS FOR NONDWELLING RECEPTACLE LOADS

Receptacle Load	Demand Factor
First 10 kVA or less	100%
Remainder over 10 kVA	50%

DEMAND FACTORS FOR KITCHEN EQUIPMENT IN NONDWELLING APPLICATIONS

Equipment	Demand Factor
1	100%
2	100%
3	90%
4	80%
5	70%
6 and over	65%

LIGHTING LOAD DEMAND FACTORS

Various Occupancy Types	Portion of Lighting Load Applied in Volt-Amps	Demand Factor in Percent
Dwelling units	First 3,000 or less at From 3,001 to 120,000 at Remainder over 120,000 at	100% 35% 25%
Hospitals	First 50,000 or less at Remainder over 50,000 at	40% 20%
Hotels and motels, including apartment houses without provision for cooking	First 20,000 or less at From 20,001 to 100,000 at Remainder over 100,000 at	50% 40% 30%
Warehouses (storage)	First 12,500 or less at Remainder over 12,500 at	100% 50%
All others	Total volt-amps of load	100%

LIGHTING LOADS BY OCCUPANCY

Various Occupancy Types	Volt-Amps per Square Foot
Armories	1
Banks	3½
Beauty parlors	3
Churches	1
Clubs	2
Court rooms	2
Dwelling units	3
Garages—commercial	½
Hospitals	2
Hotels and motels, including apartment houses without provision for cooking	2
Industrial commercial buildings	2
Lodge rooms	1½
Office buildings	3½
Restaurants	2
Schools	3
Stores	3
Warehouses In any of the preceding occupancies except one-family dwellings and individual units of two-family and multifamily dwellings	¼
Assembly halls and auditoriums	1
Halls, corridors, closets, stairways	½
Storage spaces	¼

CHAPTER 4
Receptacles, Switches, Interior Wiring, and Lighting

RECEPTACLES IN DAMP LOCATIONS

1) All receptacles installed outdoors in locations protected from the weather or in damp locations shall have weatherproof enclosures when the receptacles are covered (attached plug caps not inserted and receptacle covers closed).
2) Installations suitable for wet locations shall also be considered suitable for damp locations.
3) A receptacle shall be considered to be in a location protected from the weather where located under roofed open porches, canopies, etc., and will not be subjected to driving rain or water runoff.
4) All 15 and 20 amp, 125 and 250 volt nonlocking receptacles shall be a listed weather-resistant type and will be identified as 5-15, 5-20, 6-15, and 6-20 by the National Electrical Manufacturers Association (NEMA).

RECEPTACLES IN WET LOCATIONS

1) All 15 and 20 amp, 125 and 250 volt receptacles installed in a wet location shall have an enclosure that is weatherproof whether or not the attachment plug cap is inserted.
2) All 15 and 20 amp, 125 and 250 volt nonlocking receptacles shall be a listed weather-resistant type and will be identified as 5-15, 5-20, 6-15, and 6-20 by NEMA.
3) Receptacles installed in a wet location and subject to routine high-pressure spray washing shall be permitted to have an enclosure that is weatherproof when the attachment plug is removed.

RECEPTACLES IN WET LOCATIONS *(cont.)*

4) A receptacle installed in a wet location where the equipment to be plugged into it is not attended while in use shall have an enclosure that is weatherproof with the attachment plug cap inserted or removed.

5) A receptacle installed in a wet location where the equipment to be plugged into it will be attended while in use shall have an enclosure that is weatherproof when the attachment plug is removed.

6) Receptacles shall not be installed within or directly over a bathtub or shower stall.

7) There shall be proper protection provided for floor receptacles so that floor-cleaning equipment may be operated without damage to such receptacles.

8) The enclosure for a receptacle installed in an outlet box flush-mounted in a finished surface is made weatherproof by installing a weatherproof faceplate that provides a watertight connection between the plate and the finished surface.

TAMPER-RESISTANT RECEPTACLES IN DWELLING UNITS

All 15 and 20 amp, 125 volt receptacles shall be listed as tamper-resistant when they are installed in dwelling units.

ARC-FAULT CIRCUIT-INTERRUPTER PROTECTION

An AFCI receptacle is a device that is intended to provide personal protection from the effects of arc faults by recognizing characteristics unique to arcing and then by de-energizing the circuit when an arc fault is detected.

AFCI RECEPTACLE LOCATIONS

Residential:

Family rooms	Dens	Hallways
Dining rooms	Bedrooms	Similar rooms/areas
Living rooms	Sunrooms	Kitchens
Parlors	Rec rooms	Laundry areas
Libraries	Closets	

GFCI RECEPTACLE LOCATIONS

Residential:

Bathrooms
Boathouse hoists
Crawl spaces
Dishwasher circuits
Garages and accessory buildings
Kitchen countertops

Laundry, utility, and wet
 bar sinks
Laundry rooms
Outdoors
Swimming pools
Unfinished basements
Whirlpools and hot tubs

Nonresidential:

Bathrooms
Carnivals, circuses, and fairs
Commercial garages
Construction sites
Commercial and institutional
 kitchens
Dormitories
Locker rooms

Marinas
Outdoor public spaces
Rooftops
Swimming pools
Temporary installations
Trailer and RV parks
Vending machines
Whirlpools and hot tubs

**For Replacing Equipment
in Ungrounded Outlets:**

Light fixtures
Receptacles
Toggle switches

GFCI Notes

- A bathroom is defined as an area including a basin with a toilet, tub, or shower.

- Reference the *NEC*® for the definition of commercial and institutional kitchens.

- For applications other than light fixtures, receptacles, and switches, GFCI protection is never intended to be a substitute for grounding.

GFCI WIRING DIAGRAMS

Feed-Thru Installation

To protect the entire branch circuit, the GFCI must be the first receptacle from the circuit breaker or fuse box. Receptacles on the circuit downstream from the GFCI will also be protected.

Non-Feed-Thru Installation on a 2-Wire Circuit

Terminal protection can be achieved on a multi-outlet circuit by connecting the hot and neutral line conductors to the corresponding line side terminals of the GFCI. Only the GFCI receptacle will be protected.

WIRING DIAGRAMS FOR NEMA PLUG AND RECEPTACLE CONFIGURATIONS

2-Pole, 2-Wire Non-Grounding

125 V

1-15R   L1-15R   ML-1R

250 V

2-20R   2-30R   L2-20R

3-Pole, 3-Wire Non-grounding

125/250 V

10-20R   10-30R   10-50R   L10-20R   L10-30R   ML-3R

3φ 250 V

11-15R   11-20R   11-30R   11-50R   L11-15R   L11-20R   L11-30R

WIRING DIAGRAMS FOR NEMA PLUG AND RECEPTACLE CONFIGURATIONS *(cont.)*

2-Pole, 3-Wire Grounding

125 V

5-15R 5-20R 5-30R 5-50R L5-15R L5-20R L5-30R ML-2R

250 V

6-15R 6-20R 6-30R 6-50R L6-15R L6-20R L6-30R

277 VAC

7-15R 7-20R 7-30R 7-50R L7-15R L7-20R L7-30R

480 VAC

L8-20R L8-30R

4-6

WIRING DIAGRAMS FOR NEMA PLUG AND RECEPTACLE CONFIGURATIONS (cont.)

3-Pole, 4-Wire Grounding
125 V/250 V

14-15R 14-20R 14-30R 14-50R 14-60R L14-20R L14-30R

3φ 250 V

15-15R 15-20R 15-30R 15-50R 15-60R L15-20R L15-30R

3φ 480 V

L16-20R L16-30R

3φ 600 V

L17-30R

WIRING DIAGRAMS FOR NEMA PLUG AND RECEPTACLE CONFIGURATIONS *(cont.)*

4-Pole, 4-Wire Non-Grounding
3ϕ Wye 120 V/208 V

18-15R 18-20R 18-30R 18-50R 18-60R L18-20R L18-30R

3ϕ Wye 277 V/480 V

L19-20R L19-30R

3ϕ Wye 347 V/600 V

L20-20R L20-30R

WIRING DIAGRAMS FOR NEMA PLUG AND RECEPTACLE CONFIGURATIONS *(cont.)*

4-Pole, 5-Wire Grounding

3φ Wye 120 V/208 V

L21-20R   L21-30R

3φ Wye 277 V/480 V

L22-20R   L22-30R

3φ Wye 347 V/600 V

L23-20R   L23-30R

WIRING DIAGRAMS FOR AC SWITCHES

Single pole   Double pole

Hot
Neutral
Hot

3-way to 3-way

Hot
Neutral

3-way   4-way   3-way

Hot
Neutral

WIRING DIAGRAMS FOR PILOT LIGHT AND LIGHTED TOGGLE SWITCHES

Single-Pole Pilot Light Switch
Toggle glows when light is on.

Hot

Neutral

Single-Pole Lighted Toggle Switch
Toggle glows when switch is off.

Hot

Neutral

Double-Pole Pilot Light Switch
Toggle glows when switch is on.

Hot

Neutral

Hot

FAMILY ROOM WITH SPLIT-WIRED RECEPTACLES AND SWITCHED CIRCUIT

A split-wired receptacle has the tab between the brass (hot) terminals removed but silver (neutral) terminals remain intact. This provides either a switched circuit or two separate circuits at the same receptacle outlet.

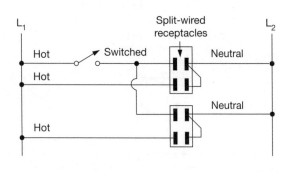

RESIDENTIAL SMOKE ALARM GUIDELINES

New Construction:
- One detector in each bedroom or room with adaptable sleeping furniture.
- One in the hall outside each sleeping area.
- One per floor, including the basement or below grade recreation room or area.
- In basements, locate near the stairs, on the underside of a ceiling joist.
- Wire to a 120 volt power supply circuit and utilize backup batteries.
- Interconnect all detectors with a 3-wire circuit with ground using NM cable.

Existing Construction:
- Add smoke detectors to existing houses when a permit is pulled for work inside the house.
- Install detectors in the same areas that are required for new construction.
- Battery-powered units may be used where you are unable to wire to the location of the smoke detector.

Do Not Install:
- Within 4" of the junction between wall and ceiling (dead air space).
- Within 3' of doorway to bathrooms and kitchens.
- Within 3' of air supply vents and the tips of paddle fans.
- Within 20' of a cooking appliance (unless the alarm has a push or silence button, or if photoelectric type).
- In dusty attics that get too hot or cold.
- Where protected by a GFCI.

RESIDENTIAL AREA REQUIREMENTS
FOR SMOKE ALARM

Number	Area Requirement
1	Smoke alarms must be located in each sleeping room.
2	Smoke alarms must be located outside of each sleeping area.
3	Smoke alarms must be located on each level of the dwelling unit including basements, but NOT including crawl spaces or uninhabitable attics.

SMOKE ALARM WIRING REQUIREMENTS

- 120 volt power supply from AFCI-protected circuit.
- Smoke alarms must be interconnected so that actuation of one alarm will activate alarms on all smoke alarms.
- Must receive primary power from building wiring (permanent) with battery backup.
- No disconnecting switch other than a circuit breaker.

FAMILY ROOM AND BEDROOM RECEPTACLE OUTLET SPACING

(A) No point along floor line shall be more than 6' from a receptacle outlet.

(B) All wall spaces 2' or larger shall have a receptacle outlet.

4-15

FAMILY ROOM WIRING DETAILS

- Receptacle outlets must be spaced so that no point along the wall is farther than 6' from the nearest outlet. A fireplace, doorway, or similar opening is not required to be counted as wall space, but receptacles must be located within 6' of each side of these spaces.
- Wall space 2' or greater must have a receptacle outlet.
- Fixed panels such as the portion of a sliding glass door that is not movable must be considered wall space.
- Switched lighting outlet is required. Lighting fixture or switched plug is acceptable.
- GFCI protection required for all 120 volt, 15 or 20 amp receptacles located within 6' of a wet bar sink.
- If wall space is not available, receptacles must be installed on the floor. Receptacles installed on the floor must be within 18" of the wall to be counted as a required receptacle.
- All 15 and 20 amp, 120 volt receptacle outlets must be listed as tamper-resistant.
- Arc-fault protection is also required for all outlets including all lighting fixtures and receptacles for family rooms including wetbar areas, bonus rooms, and living rooms.

WETBAR AND TRACK LIGHTING WIRING DETAILS

- GFCI protection required for all 120 volt, 15 or 20 amp receptacles located within 6' of a wetbar sink and there are no exceptions for single outlet or dedicated appliances.
- GFCI protection is NOT required for receptacles located 6' or further from the outside edge of a wetbar sink, but they must be AFCI protected and listed tamper-resistant.
- Space occupied by a countertop must be counted as wall space and satisfy the 6' requirement for placing receptacles along the wall.
- Track lighting is not allowed less than 5' above the floor unless protected from possible damage.
- Track lighting must be hardwired and must be supported at a minimum of two points every 4'.

BEDROOM WIRING DETAILS

- Receptacle outlets must be spaced so that no point along the wall is farther than 6' from the nearest outlet.
- Wall space 2' or greater must have a receptacle outlet.
- Switched lighting outlet is required. Lighting fixture or switched plug is acceptable.
- Smoke detector required inside all sleeping rooms.
- Arc-fault protection required for all outlets, which includes all lighting fixtures, smoke detectors, and receptacles.
- All 15 and 20 amp, 120 volt receptacle outlets must be listed tamper-resistant.

CLOTHES CLOSET WIRING DETAILS

- Open bulb incandescent lighting fixtures are not allowed.
- Storage space is considered 12" (minimum) from side and rear walls or the shelf width, whichever is greater.
- Recessed and enclosed incandescent and fluorescent lights must be 6" from nearest storage space and 18" from side and rear walls.
- Surface-mounted incandescent lights must be 12" from nearest storage space and 24" from side and rear walls.
- Surface-mounted fluorescent lights must be 6" from nearest storage space and 18" from side and rear walls.
- All 15 and 20 amp, 120 volt outlets must be AFCI protected and listed tamper-resistant.
- Lighting and receptacle outlets are not required unless it is a walk-in closet.

HALLWAY AND STAIRWELL WIRING DETAILS

- Wall switch controlled lighting outlet required.
- Where there are 6 or more risers on the stairs, a 3 way or 4 way wall switch to control the lighting outlet is required at each level.
- Hallways 10' or longer must have a receptacle outlet.
- Smoke detectors must be located outside of each sleeping area.
- All 15 and 20 amp, 120 volt outlets must be AFCI protected and listed tamper-resistant.

LAUNDRY ROOM WIRING DETAILS

- Switched lighting outlet required.
- Dedicated 20 amp circuit required for laundry receptacle outlets. No other devices allowed. At least 1 receptacle outlet required for laundry purposes.
- GFCI protection required for all 120 volt, 15 or 20 amp receptacles.
- 4-wire, 30 amp circuit required for electric dryers.
- All 15 and 20 amp, 120 volt receptacle outlets must be listed tamper resistant.
- Clothes dryers come from the factory with a bonding strap joining the ground and neutral terminals. This strap must be removed to comply with the code. Follow manufacturer's instructions to make cord connections.

KITCHEN RECEPTACLES

All receptacles must be on small appliance branch circuits. Both AFCI and GFCI protection are required for all receptacles serving kitchen areas.

COUNTERTOP RECEPTACLE OUTLET SPACING

Maximum 18"   AFCI/GFCI   AFCI/GFCI

24" or greater

Minimum one outlet

ISLAND

12" or greater

AFCI/GFCI

12" or less below countertop

12" or greater

Minimum one outlet, or wall receptacle within 6 ft.

Countertop extension maximum 6"

24" or greater

AFCI/GFCI   Within 4'   12" or wider countertops   AFCI/GFCI

Countertop may not extend more than 2' from an outlet.

KITCHEN WIRING DETAILS

- No point along the countertop is further than 2' from a receptacle and counter spaces longer than 1' require a receptacle.
- One wall switch controlled lighting fixture required.
- If spacing behind the sink is less than 12", one receptacle is required within 2' of each side.
- If spacing behind the sink is 12" or greater, this area must be counted as counter space.
- Counter spaces divided by range tops, refrigerators, or sinks must be considered separate spaces.
- An island or peninsula space requires a receptacle only if larger than 24" × 12". If a sink or cooktop is installed on the island, and the width of the counter behind the appliance is less than 12", a receptacle must be installed on both sides. (A countertop wall receptacle within 6 ft. is acceptable).
- If spacing behind cooktop is less than 18 in. one receptacle is required within 2' of each side.
- If spacing behind cooktop is 18" or greater, this area must be counted as wall space.
- No point along the wall may be further than 6' from a receptacle, if not counter space.
- All 15 and 20 amp, 120 volt receptacle outlets in kitchens must be listed tamper-resistant.
- All 15 and 20 amp, 120 volt receptacle outlets in dining rooms and eating areas must be AFCI protected and listed tamper-resistant.
- Receptacles must be AFCI and GFCI protected.
- All receptacles in kitchen area must be on a 20 amp circuit.
- Minimum of two 20 amp circuits required to supply kitchen counter top receptacles.

KITCHEN WIRING DETAILS *(cont.)*

- Receptacle for refrigerator should be supplied from a dedicated 15 or 20 amp circuit.
- Garbage disposal flex cord (use 3 wire type) can be no shorter than 18" and no longer than 36" and can be hard wired.
- Disposal receptacle must be AFCI/GFCI protected and located in the space occupied by the dishwasher or adjacent cabinet.
- Switch is allowed for disposal disconnect if not cord-and-plug connected.
- Length of dishwasher cord (use 3 wire type) must extend 3'–4' from the rear of the appliance. Dishwasher is normally hard wired.
- Receptacles for microwaves should be mounted in the cabinet directly above the appliance. GFCI and AFCI protection are required. Microwaves are normally supplied by a dedicated 20 amp circuit, 12/2 with ground, copper, nonmetallic cable.
- Ranges rated 8¾ kW or more must be supplied by a minimum 40 amp circuit.
- Receptacles to be mounted no more than 18" above the countertop.
- 4-wire circuit required for cord-and-plug connected free standing ranges which are typically wired with 50 amp circuit, 6/3 with ground, copper, nonmetallic cable.
- Range hoods and built-in microwaves may be cord-and-plug connected if the receptacle is accessible and supplied from an individual circuit.

BATHROOM RECEPTACLES AND SPACING

At least one receptacle outlet shall be installed within 36" of outside edge of each basin and on a wall adjacent to basin location.

All GFCI-protected receptacles shall be supplied by at least one 20 amp circuit with no other type of outlets on the circuit.

All bathroom receptacles in dwelling units are required to be GFCI-protected.

BATHROOM WIRING DETAILS

- One wall switch controlled lighting outlet required.
- All receptacles must be GFCI protected and wired on a 20 amp circuit and at least one receptacle must be installed within 3' of each sink.
- No other devices such as lights or vent fans are allowed on the receptacle circuit, unless the dedicated 20 amp circuit supplies only one bathroom.
- Different bathroom receptacles may be served from one 20 amp circuit, but only those receptacles, not any other devices.
- No part of a lighting track, hanging lighting fixture or ceiling fan allowed within a 3' horizontal by 8' vertical zone above the threshhold of a shower or the rim of a bathtub. Recessed or surface mounted fixtures are allowed, but must be listed for damp locations.
- Receptacles or switches are not allowed within or directly over a tub or shower space.
- Access is required to all electrical equipment for hydro-massage bathtubs.
- Hydromassage bathtub electrical equipment must be GFCI protected and wired on a dedicated 20 amp circuit.
- Metal piping systems must be bonded to pump motor with #8 solid copper wire.
- All 15 and 20 amp, 120 volt receptacle outlets must be listed tamper resistant.
- Vent fans installed directly above showers must be GFCI protected.

CRAWL SPACE AND INSIDE EQUIPMENT WIRING DETAILS

- Receptacle required if heating or air conditioning equipment is present. Receptacle must be located on the same level and within 25' of the equipment.
- Receptacles must be GFCI protected.
- Lighting outlet required if space is used for storage or contains equipment that may need servicing and must be located at or near the equipment (HVAC, etc.).
- Switch for lighting outlet must be located at the point of entrance to the storage or equipment space.
- Disconnecting means for HVAC equipment must be located within sight and readily accessible from the equipment. The disconnect may not be mounted on the panels designed to allow access to the equipment.
- All 15 and 20 amp, 120 volt receptacle outlets in dwellings must be listed tamper-resistant.
- All 15 and 20 amp, 120 and 240 volt nonlocking receptacles in wet or damp locations must be listed weather-resistant type.
- Cables containing wires smaller than two #6 or three #8 are not allowed to be secured directly to the bottom of the ceiling joists. These wires must be routed through bored holes, installed on running boards or through conduit raceways.
- The access must be sized large enough to remove the largest piece of equipment out of the space.
- The passageway to the equipment must be floored and not less than 24" wide or wider to allow equipment installation or removal.
- A 30" × 30" level space must be provided in front of the equipment for servicing.

UNFINISHED BASEMENT WIRING DETAILS

- Wall switch controlled lighting outlet is required.
- Switched lighting outlet required on exterior to illuminate all personnel entrances.
- At least one receptacle outlet is required in unfinished basements.
- Where a portion of the basement is finished into a habitable room, one receptacle outlet is required in each unfinished portion.
- All 15 and 20 amp, 120 volt receptacle outlets must be listed tamper-resistant.
- GFCI protection is required for all 15 and 20 amp, 120 volt receptacles.
- GFCI protection is required for all 15 and 20 amp, 120 volt receptacles that supply sump pumps.
- GFCI protection required if receptacles are not readily accessible such as ceiling receptacles for sump pump alarms.
- GFCI protection required for receptacles that are dedicated for appliances that are not easily moved such as a refrigerator, freezer or washer.
- 15 and 20 amp, 120 volt receptacles must be GFCI protected within 6' of a utility sink.
- Exception: GFCI protection not required for a receptacle supplying a permanently installed burglar or fire alarm system.
- Smoke alarm must be installed for the basement level.
- NM cable must closely follow the surface of the building or be protected from damage. (Must protect horizontal runs of exposed cable.)
- NM cable used on the wall of an unfinished basement is permitted to be installed in listed conduit. (Must use nonmetallic bushing at the point where cable enters.)
- Cables containing wires smaller than two #6 or three #8 are not allowed to be secured directly to the bottom of the ceiling joists. (Must be routed through bored holes, installed on running boards or in conduit raceways.)

GARAGE AND SHOP RECEPTACLES

Dedicated appliance

GFCI protection is required for all receptacles in a garage not intended as a living space.

GFCI protection is required for dedicated receptacles.

Grade level

Accessory buildings at or below grade not intended as living space require GFCI protection on all receptacles.

GARAGE AND SHOP WIRING DETAILS

- Wall switch controlled lighting outlet is required.
- Switched lighting outlet, required on exterior to illuminate all personnel entrances, but is not required for vehicle doors, such as overhead garage doors.
- At least one receptacle outlet is required in attached garages, detached garages and shops.
- GFCI protection is required for all 15 and 20 amp, 120 volt receptacles.
- GFCI protection required if receptacles are not readily accessible such as a ceiling receptacle for a garage door opener.
- GFCI protection required for receptacles that are dedicated for appliances that are not easily moved such as a refrigerator, freezer or a washer.
- 15 and 20 amp, 120 volt receptacles must be GFCI protected within 6' of a utility sink.
- All 15 and 20 amp, 120 volt receptacle outlets must be listed tamper-resistant.
- Some jurisdictions require smoke detectors and/or carbon monoxide detectors for garages. Always check with your local building codes or the electrical inspection department.

STORAGE AREA WIRING DETAILS

- Switched lighting outlet is required and must be located at the point of entrance to the space.
- All 15 and 20 amp, 120 volt receptacle outlets must be listed tamper-resistant.
- Receptacle outlet required for HVAC equipment.
- Smoke detectors not required in uninhabitable attics.

OUTDOOR RECEPTACLES

GFCI protection is not required for receptacles utilizing snow and ice melting equipment and are not readily accessible.

A GFCI receptacle is required for all outdoor plug in lighting.

A GFCI receptacle is required regardless of height except as noted above.

GFCI protection is required on all outdoor receptacles on dwelling units.

OUTDOOR WIRING DETAILS

- Receptacle outlet required at the front and rear located no higher than 6' 6" above grade.
- All 15 and 20 amp, 120 volt receptacles must be GFCI protected.
- Receptacles installed in damp locations must have a weather resistant cover.
- Receptacles installed in wet locations must have a cover that is weatherproof whether or not a cord is plugged in.
- Switched lighting outlet required to illuminate all personnel entrances.
- All 15 and 20 amp, 120 volt receptacle outlets must be listed tamper-resistant.
- All 15 and 20 amp, 120 and 240 volt nonlocking receptacles in wet or damp locations must be listed weather-resistant-type.
- HVAC equipment must have receptacle on the same level and located within 25'.
- Disconnecting means for HVAC equipment must be located within sight and readily accessible from the equipment.
- Must maintain clear working space at all electrical equipment that may require servicing while energized.
- Conduit installed outside in a wet location must be listed for use outdoors.
- Conductors installed outdoors must be listed for use in wet locations.

RECOMMENDED LIGHT LEVELS

Interior		Exterior	
Area	**fc**	**Area**	**fc**
Assembly		**Airports**	
Rough	30	Terminal	10
Medium	100	Loading	2
Banks		**Buildings**	
Lobby	50	Light surface	15
Tellers	150	Dark surface	50
Hospital/Med.		**Construction**	
Dental	1,000	General	10
Operating	2,500	Excavation	2
Machine Shop		**Parking Areas**	
Rough	50	Industrial	2
Medium	100	Shopping	5
Offices		**Loading Areas**	
Regular	100	Pier	20
Detailed work	200	Trucking	30
Printing		**Service Station**	
Proofreading	150	Pumps	25
Color inspecting	200	Service	5

LAMP ADVANTAGES AND DISADVANTAGES

Lamp	Advantages	Disadvantages
Incandescent, tungsten-halogen	Low initial cost Simple construction No ballast required Available in many shapes and sizes Requires no warm-up or restart time Inexpensively dimmed Simple maintenance	Low electrical efficiency High operating temperature Short life Bright light source in small space Does not allow large distribution of light
Fluorescent	Available in many shapes and sizes Moderate cost Good electrical efficiency Long life Low shadowing Low operating temperature Short turn-ON delay	Not suited for high-level light in small, highly-concentrated applications Requires ballast Higher initial cost than incandescent lamps Light output and color affected by ambient temperature Expensive to dim
Low-pressure sodium, mercury-vapor, metal-halide, high-pressure	Good electrical efficiency Long life High light output Slightly affected by ambient temperature	May cause color distortion Long start and restart time High initial cost High replacement cost Requires ballast Expensive or not possible to dim Problem starting in cold weather High-socket voltage required

LAMP RATINGS

Lamp (W = Watts)	Initial Lumen	Mean Lumen
40 W standard incandescent	480	N/A
100 W standard incandescent	1,750	N/A
40 W standard fluorescent	3,400	3,100
100 W tungsten-halogen	1,800	1,675
100 W mercury-vapor	4,000	3,000
250 W mercury-vapor	12,000	9,800
100 W high-pressure sodium	9,500	8,500
250 W high-pressure sodium	30,000	27,000
250 W metal-halide	20,000	17,000

LUMENS, LIFE, AND EFFICACY FOR VARIOUS LAMPS

Lamp Type (watts)	Initial Lumens	Life (hours)	Efficacy (lumens/watts)
Incandescent, standard inside frosted			
25	235	1,000	9
40	480	1,500	12
60	840	1,000	14
75	1,210	850	16
100	1,670	750	17
150	2,850	750	19
200	3,900	750	19
300	6,300	1,000	21
500	10,750	1,000	21
1,000	23,100	1,000	23
Incandescent PAR-38			
150 Spot	1,100	2,000	7
150 Flood	1,350	2,000	9
Quartz incandescent			
500	10,550	2,000	21
1,000	21,400	2,000	21
1,500	35,800	2,000	24
40	3,150	20,000	78
40	2,200	20,000	55
60	4,300	12,000	41
60	3,050	12,000	50
85	2,850	20,000	81
85	2,000	20,000	57
Fluorescent, U-line			
40	2,850	12,000	71
40	2,020	12,000	50
Fluorescent, instant start			
60	5,600	12,000	93
60	4,000	12,000	66
75	6,300	12,000	84
75	4,500	12,000	60
95	8,500	12,000	89
95	6,100	12,000	64
110	9,200	12,000	83
110	6,550	12,000	59
180	12,300	10,000	68
215	14,500	10,000	67
215	13,600	10,000	63

LUMENS, LIFE, AND EFFICACY FOR VARIOUS LAMPS *(cont.)*

Lamp Type (watts)	Initial Lumens	Life (hours)	Efficacy (lumens/watts)
Mercury-vapor			
40	1,140	24,000	28
50	1,575	24,000	31
75	2,800	24,000	37
100	4,300	24,000	43
175	8,500	24,000	48
175	7,900	24,000	45
250	13,000	24,000	52
250	12,100	24,000	48
400	23,000	24,000	57
400	21,000	24,000	52
700	43,000	24,000	61
1,000	63,000	24,000	63
Metal-halide			
175	14,000	7,500	80
250	20,500	7,500	82
400	34,000	15,000	85
1,000	105,000	10,000	105
High-pressure sodium			
50	3,300	20,000	66
70	5,800	20,000	82
100	9,500	20,000	95
150	16,000	24,000	106
150	12,000	12,000	80
200	22,000	24,000	110
215	19,000	12,000	88
250	27,500	24,000	110
400	50,000	24,000	125
1,000	140,000	24,000	140
Low-pressure sodium			
65	4,800	18,000	137
65	8,000	18,000	145
90	13,500	18,000	150
135	22,500	18,000	166
180	33,000	18,000	183

LIGHT SOURCE CHARACTERISTICS

Type of Characteristic	Incandescent, Including Tungsten	Fluorescent	High-Intensity Discharge			
			Mercury-Vapor (Self-ballasted)	Metal-Halide	High-Pressure Sodium (Improved Color)	Low-Pressure Sodium
Wattages (lamp only)	15–1,500	15–219	40–1000	175–1000	70–1000	35–180
Life[a] (hr)	750–12,000	7,500–24,000	16,000–15,000	1,500–15,000	24,000 (10,000)	18,000
Efficacy[a] (lumens/W) lamp only	15–25	55–100	50–60 (20–25)	80–100	75–140 (67–112)	Up to 180
Lumen maintenance	Fair to excellent	Fair to excellent	Very good (good)	Good	Excellent	Excellent
Color rendition	Excellent	Good to excellent	Poor to excellent	Very good	Fair	Good
Light direction control	Very good to excellent	Fair	Very good	Very good	Very good	Fair
Source size	Compact	Extended	Compact	Compact	Compact	Extended
Relight time	Immediate	Immediate	3–10 min	10–20 min	Less than 1 min	Immediate
Comparative fixture cost	Low: simple fixtures	Moderate	Higher than incandescent and fluorescent	Generally higher than mercury	High	High
Comparative operating cost	High: short life and low efficiency	Lower than incandescent	Lower than incandescent	Lower than mercury	Lowest of HID types	Low
Auxiliary equipment needed	Not needed	Needed: medium cost	Needed: high cost	Needed: high cost	Needed: high cost	Needed: high cost

[a]Life and efficacy ratings subject to revision. Check manufacturers' data for latest information.

4-35

COMMON FLUORESCENT TYPES

Type	Watts	Length	Size	Base
T-5	4–13	6–21	⅝	Miniature Bi-pin
T-5	13	21	⅝	Miniature Bi-pin
T-8	15–30	18–36	1	Medium Bi-pin
T-9	20–40	6–16	1⅛	Four-pin
T-10	40	48	1¼	Medium Bi-pin
T-12	14–75	15–96	1½	Medium Bi-pin
T-12	60–75	96	1½	Single-Pin

Note: All lengths and sizes are in inches.

COMMON SHAPE IDENTIFIERS

C – Circular lamp. The letter "C" typically appears in the prefix, immediately after the "F."

T – Tubular lamp, omitted from the F40 lamp code for rapid-start lamps.

J – Used in conjunction with a "T" designator, "J" denotes a lamp with a clear outer glass jacket over the regular "T" envelope, which acts as a thermal insulator and permits operation of the lamps at subzero temperatures.

BX, **PL**, and **PTT**
  – These identifiers denote a family of single-ended compact lamps.

U – A lamp shaped like the letter "U."

CHAPTER 5
Grounding and Bonding

RESISTIVITIES OF DIFFERENT METALS

Metal Type	Ohms/ft	Metal Type	Ohms/ft
Brass	43	Aluminum	17
Copper	11	Monel	253
Silver	10	Nichrome	600
Gold	15	Nickel	947
Iron (Pure)	60	Tantalum	93
Magnesium	276	Tin	69
Manganin	265	Tungsten	34

Note: The values above represent the resistance offered by a wire 1 foot long with a diameter of 1 mil at an ambient temperature of 30°C. Resistance varies with temperature change.

RESISTIVITIES OF DIFFERENT SOILS

Soil Type	Resistivity in ohms per cubic meter		
	Average	Minimum	Maximum
Fills—ashes, cinders, brine wastes	2,370	590	7,000
Clay, shale, gumbo, loam	4,060	340	16,300
Same—with varying proportions of sand and gravel	15,800	1,020	135,000
Gravel, sand, stones, with little clay or loam	94,000	59,000	458,000

BASIC GROUNDED CONDUCTOR RULES

Circuit breakers or switches shall not disconnect the grounded conductor of a circuit.

Hot

Conductors in same raceway

3-way switch

L

3-way switch

Neutral

Do not switch the neutral conductor.

Hot

Light

Switch in neutral conductor

120 V

Neutral

Accidental ground

Connected to grounded neutral bus

Use DP switch in grounded neutral conductor.

Exception: A circuit breaker or switch may disconnect grounded circuit conductor if all circuit conductors are disconnected at the same time.

BASIC GROUNDED CONDUCTOR RULES *(cont.)*

Circuit breakers or switches shall not disconnect the grounded conductor of a circuit.

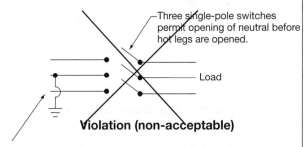

Three single-pole switches permit opening of neutral before hot legs are opened.

Load

Violation (non-acceptable)

3-wire, 1ϕ supply

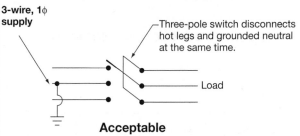

Three-pole switch disconnects hot legs and grounded neutral at the same time.

Load

Acceptable

Exception: A circuit breaker or switch may disconnect the grounded circuit conductor if it cannot be disconnected until all other ungrounded conductors have been disconnected.

GROUNDED CONDUCTOR — NEUTRAL

3ϕ, 4-Wire, Wye Grounded System

GROUNDED CONDUCTOR — NOT NEUTRAL

3ϕ, 3-Wire, Corner-Grounded Delta System

GROUNDING DIFFERENT TYPES OF CIRCUITS

Single-Phase, Two-Wire

Single-Phase, Three-Wire

Three-Phase Delta, Three-Wire

Three-Phase Delta, Four-Wire

Three-Phase Wye

208 V

208 V

208 V

208 V

120 V 120 V 120 V

Grounded conductor
white or
natural gray

4-Wire System with a Neutral

GROUNDING AN EXISTING CIRCUIT

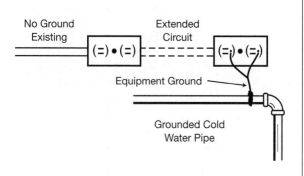

No Ground Existing

Extended Circuit

(=) • (=)

(=) • (=)

Equipment Ground

Grounded Cold Water Pipe

GROUNDING A SCREW-SHELLBASE

H

N

Hot

Screw-shell Must Be Grounded

BONDING A TYPICAL SERVICE ENTRANCE

- Grounding bushing
- Neutral
- Grounding bus
- Water pipe
- Load

- Threaded hub
- Neutral
- Service equipment
- Bonding bushing
- Meter housing
- To ground
- Load

BONDING SERVICE EQUIPMENT METHODS

Meter socket grounded conductor

Grounded Service Conductor

Internal Threadless Connections

Internal Threaded Connections

Bonding bushing

Bonding and grounding wedges

Bonding locknut

Other Devices

BONDING JUMPERS FOR GROUNDING EQUIPMENT CONDUCTOR RACEWAYS

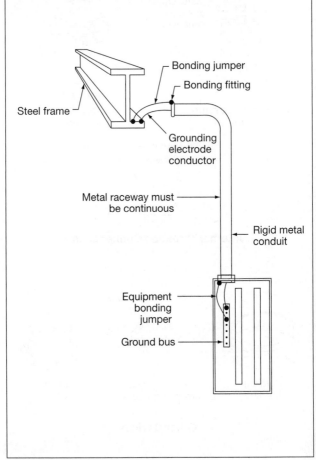

Steel frame

Bonding jumper

Bonding fitting

Grounding electrode conductor

Metal raceway must be continuous

Rigid metal conduit

Equipment bonding jumper

Ground bus

EQUIPMENT BONDING JUMPERS

Motor disconnect

Exterior equipment bonding jumper (6' maximum length if outside raceway.)

2' Flexible metallic conduit

EQUIPMENT GROUNDING CONDUCTORS

Steel frame

Typical service panel

Grounded conductor

Main bonding jumper

Grounding electrode conductor

Equipment grounding conductor

Typical motor

Ground rod

5-12

MINIMUM SIZE CONDUCTORS FOR GROUNDING RACEWAY AND EQUIPMENT

Settings of automatic overcurrent devices in circuits ahead of equipment, conduit, etc. are not to exceed the ampacity ratings below.	Conductor Size (AWG or kcmil)	
	Copper	Aluminum or copper-clad aluminum
15	14 AWG	12 AWG
20	12	10
30	10	8
40	10	8
60	10	8
100	8	6
200	6	4
300	4	2
400	3	1
500	2	1/0
600	1	2/0
800	1/0	3/0
1,000	2/0	4/0
1,200	3/0	250 kcmil
1,600	4/0	350
2,000	250 kcmil	400
2,500	350	600
3,000	400	600
4,000	500	800
5,000	700	1,200
6,000	800	1,200

BASIC GROUNDING CONNECTIONS

Service entrance raceway (utility feed)

Steel post

Grounding electrode conductor bonded to steel structural member

Typical service panel

Bonding jumper, main

Neutral bus (ground)

Grounding electrode conductor bonded to rebar ufer ground

Grounding electrode conductor

Concrete floor slab

Clamp

Raceway

Iron rebar inside slab (ufer ground)

Ground rod

GROUNDING ELECTRODE CONDUCTORS FOR AC SYSTEMS

Service-Entrance Conductor or Equivalent Area for Parallel Conductors		Grounding Electrode Conductor Size and Type	
Copper	Aluminum or Copper-Clad Aluminum	Copper	Aluminum or Copper-Clad Aluminum
2 OR SMALLER	1/0 OR SMALLER	8	6
1 OR 1/0	2/0 OR 3/0 AWG	6	4
2/0 OR 3/0 AWG	4/0 OR 250 kcmil	4	2
OVER 3/0 THRU 350 kcmil	OVER 250 THRU 500 kcmil	2	1/0
OVER 350 kcmil THRU 600 kcmil	OVER 500 kcmil THRU 900 kcmil	1/0	3/0
OVER 600 kcmil THRU 1,100 kcmil	OVER 900 kcmil THRU 1,750 kcmil	2/0	4/0
OVER 1,100 kcmil	OVER 1,750 kcmil	3/0	250 kcmil

A) The table above applies to the derived conductors of separately derived AC systems.

B) When multiple sets of service conductors are utilized, the equivalent size of the largest service-entrance conductor shall be determined by the largest sum of the areas of the corresponding conductors of each set.

C) If there are no service-entrance conductors, the grounding electrode conductor size shall be determined by the equivalent size of the largest service-entrance conductor required for the load to be served.

D) Refer to the NEC® installation restrictions concerning aluminum and copper-clad aluminum conductors.

TYPES OF GROUNDING METHODS PER THE NEC®

Below ground metal water supply pipe

10' min

Ground lug

Metal frame

Metal building frame

Concrete-encased electrode #4 or larger conductor minimum 20' in length

24" or more

Ground ring – #2 bare copper conductor minimum 20' in length

GROUNDING A BASIC NONMETALLIC UNDERGROUND SERVICE RACEWAY

Utility meter

2" PVC conduit

Finished grade

18" minimum

Service lateral

2" PVC conduit

2" metal elbow not required to be grounded

Metal elbows installed in nonmetallic raceways are not required to be grounded as long as they are buried to a minimum of 18" below the finished grade.

GROUNDING A BASIC
OVERHEAD ELECTRICAL SERVICE

Service head

Service drop

Service raceway

Service panel

Ground bus

Utility meter

Grounding electrode conductor from utility meter (if permissable)

Ground rod

Grounding electrode conductor

GENERAL REQUIREMENTS FOR GROUNDING AND BONDING SERVICES

A) All grounded AC services for premises shall have a grounding electrode conductor connected to the grounded service conductor.

B) For minimum size of service neutrals, use the largest of:
  - the maximum unbalanced loads.
  - the size of the grounding electrode conductor.

C) Size the grounding electrode conductor based on the largest service entrance conductor or reference the chart on page 5-15.

D) Size the equipment bonding jumper based on the size of the grounding electrode conductor on the supply side of the service.

E) Run a neutral to each service disconnect and bond the neutral bus to the enclosure.

F) Bond all metal service conduits to enclosures by:
  - using bonding jumpers across knockouts.
  - using bonding locknuts for cut holes and outer rings.
  - ensuring the conductor size to ground rods is 6 AWG.

G) Types of grounding electrodes:

  - Minimum 10' of metal water service pipe in soil
  - Grounded structural steel in a building
  - Re-bar or steel rod encased in concrete
  - Buried metal plates

Note: Two grounding electrodes are required.

H) When bonding to interior metal building structures use:
  - metal water pipes or other metal piping systems.
  - structural steel systems.

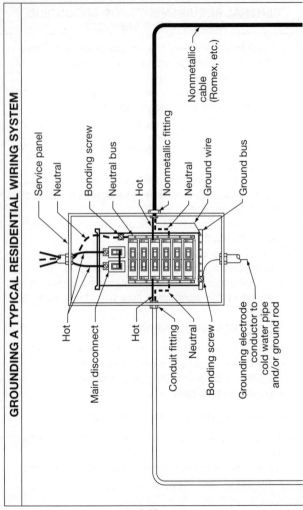

GROUNDING A TYPICAL RESIDENTIAL WIRING SYSTEM

Service panel
Neutral
Bonding screw
Neutral bus
Hot
Nonmetallic fitting
Neutral
Ground wire
Ground bus
Nonmetallic cable (Romex, etc.)

Hot
Main disconnect
Hot
Conduit fitting
Neutral
Bonding screw
Grounding electrode conductor to cold water pipe and/or ground rod

5-20

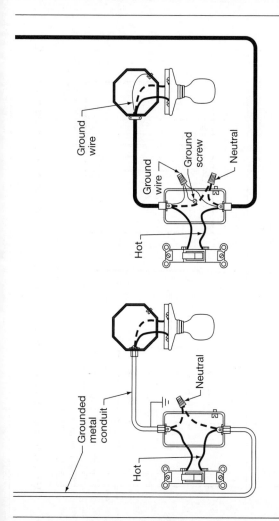

Ground wire

Ground wire

Ground screw

Neutral

Hot

Grounded metal conduit

Neutral

Hot

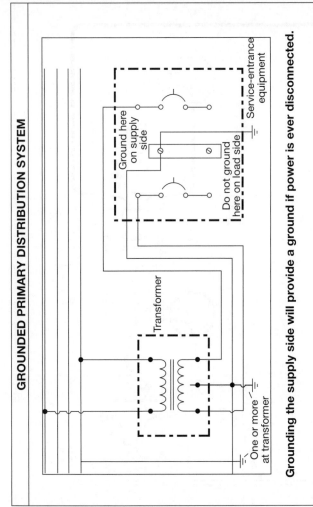

GROUNDED PRIMARY DISTRIBUTION SYSTEM

Transformer

One or more at transformer

Service-entrance equipment

Ground here on supply side

Do not ground here on load side

Grounding the supply side will provide a ground if power is ever disconnected.

CHAPTER 6
Motors

DESIGNING MOTOR BRANCH CIRCUITS

For one motor:

1. Determine full-load current of motor(s).

2. Multiply full-load current by 1.25 to determine minimum conductor ampacity.

3. Determine wire size.

4. Determine conduit size.

5. Determine minimum fuse or circuit breaker size.

6. Determine overload rating.

For more than one motor:

1. Perform steps 1 through 6 as shown above for each motor.

2. Add full-load current of all motors, plus 25% of the full-load current of the largest motor to determine minimum conductor ampacity.

3. Determine wire size.

4. Determine conduit size.

5. Add the fuse or circuit breaker size of the largest motor, plus the full-load currents of all other motors to determine the maximum fuse or circuit breaker size for the feeder.

MOTOR BRANCH CIRCUIT REQUIREMENTS

Branch Circuit Conductors:

In general, the conductors that supply a single motor used in a continuous duty application shall have an ampacity of not less than 125% of the motor's highest full-load current rating. For multiple motors add the full-load current of all the motors plus 25% of the highest full-load current of the largest motor.

Motor Disconnecting Means:

An individual disconnecting means shall be provided for each motor controller and must be in sight of both the motor and controller.

Fuses or Circuit Breakers:

Time-delay fuses for most motors are rated at 175% of the full-load amps.

Circuit breakers for most motors are rated at 250% of the full-load amps.

Note: Refer to the chart on overcurrent protective device ratings on page 6-14 for additional information.

Motor Controllers:

The motor controller shall have the same horsepower rating as the motor at the application voltage. A branch-circuit inverse time circuit breaker and a molded case switch shall be permitted as controllers.

Motor Overloads:

The overload device shall be selected to trip at:

125% — For motors with a service factor of 1.15 or greater or motors with a marked temperature rise of 40°C or less.

115% — For all other motors.

A thermally protected motor does not require additional overload protection.

Motor Full-Load Currents (Ampacities):

To size overload protective devices, use the full-load amps rating from the motor nameplate or refer to the charts on motor full-load current ratings located in this chapter on pages 6-5 through 6-9.

NEMA RATINGS OF AC MOTOR STARTERS IN AMPS PER HP, PHASE AND VOLTAGE

Nema Motor Str. Size	8 hr Open Rating in AMPS	Motor Horsepower				
		3ϕ			1ϕ	
		200 V (208)	230 V (240)	460/575 V (480/610)	115 V (120)	230 V (240)
00	9	1½	1½	2	⅓	1
0	18	3	3	5	1	2
1	27	7½	7½	10	2	3
2	45	10	15	25	3	7½
3	90	25	30	50	7½	15
4	135	40	50	100	—	—
5	270	75	100	200	—	—
6	540	150	200	400	—	—
7	810	—	300	600	—	—
8	1,215	—	450	900	—	—
9	2,250	—	800	1,600	—	—

FULL-LOAD CURRENTS IN AMPS FOR VARIOUS THREE-PHASE AC MOTORS

HP	Induction Type Squirrel-Cage and Wound-Rotor							Synchronous Type*			
	115 V	200 V	208 V	230 V	460 V	575 V	2300 V	230 V	460 V	575 V	2300 V
½	4.4	2.5	2.4	2.2	1.1	0.9	—	—	—	—	—
¾	6.4	3.7	3.5	3.2	1.6	1.3	—	—	—	—	—
1	8.4	4.8	4.6	4.2	2.1	1.7	—	—	—	—	—
1½	12.0	6.9	6.6	6.0	3.0	2.4	—	—	—	—	—
2	13.6	7.8	7.5	6.8	3.4	2.7	—	—	—	—	—
3	—	11.0	10.6	9.6	4.8	3.9	—	—	—	—	—
5	—	17.5	16.7	15.2	7.6	6.1	—	—	—	—	—
7½	—	25.3	24.2	22.0	11.0	9.0	—	—	—	—	—
10	—	32.2	30.8	28.0	14.0	11.0	—	—	—	—	—
15	—	48.3	46.2	42.0	21.0	17.0	—	—	—	—	—
20	—	62.1	59.4	54.0	27.0	22.0	—	—	—	—	—
25	—	78.2	74.8	68.0	34.0	27.0	—	53.0	26.0	21.0	—
30	—	92.0	88.0	80.0	40.0	32.0	—	63.0	32.0	26.0	—
40	—	120	114	104	52.0	41.0	—	83.0	41.0	33.0	—
50	—	150	143	130	65.0	52.0	—	104	52.0	42.0	—
60	—	177	169	154	77.0	62.0	16.0	123	61.0	49.0	12.0
75	—	221	211	192	96.0	77.0	20.0	155	78.0	62.0	15.0
100	—	285	273	248	124	99.0	26.0	202	101	81.0	20.0
125	—	359	343	312	156	125	31.0	253	126	101	25.0
150	—	414	396	360	180	144	37.0	302	151	121	30.0
200	—	552	528	480	240	192	49.0	400	201	161	40.0
250	—	—	—	—	302	242	60.0	—	—	—	—
300	—	—	—	—	361	289	72.0	—	—	—	—
350	—	—	—	—	414	336	83.0	—	—	—	—
400	—	—	—	—	477	382	95.0	—	—	—	—
450	—	—	—	—	515	412	103	—	—	—	—
500	—	—	—	—	590	472	118	—	—	—	—

Amperages listed are permitted for system voltage ranges of 110–120, 220–240, 440–480, and 550–600 volts.
*For 90% power factor, multiply the amps by 1.10.
*For 80% power factor, multiply the amps by 1.25.

THREE-PHASE AC MOTOR REQUIREMENTS

Motor HP	Rated Volts	FLC Amps	OCPD Size	Starter Size	Heater Amps	Wire Size	Conduit Size
½	230	2.2	15	00	2.530	12	¾"
	460	1.1	15	00	1.265	12	¾"
¾	230	3.2	15	00	3.680	12	¾"
	460	1.6	15	00	1.840	12	¾"
1	230	4.2	15	00	4.830	12	¾"
	460	2.1	15	00	2.415	12	¾"
1½	230	6.0	15	00	6.90	12	¾"
	460	3.0	15	00	3.45	12	¾"
2	230	6.8	15	0	7.82	12	¾"
	460	3.4	15	00	3.91	12	¾"
3	230	9.6	20	0	11.04	12	¾"
	460	4.8	15	0	5.52	12	¾"
5	230	15.2	30	1	17.48	12	¾"
	460	7.6	15	0	8.74	12	¾"
7½	230	22.0	45	1	25.30	10	¾"
	460	11.0	20	1	12.65	12	¾"
10	230	28.0	60	2	32.20	10	¾"
	460	14.0	30	1	16.10	12	¾"
15	230	42.0	70	2	48.30	6	1"
	460	21.0	40	2	24.15	10	¾"

Notes:
1) Overcurrent device may be increased due to starting current and load conditions.
2) Wire size based on 75°C.
3) Overload heater based on motor nameplate.
4) Conduit size based on rigid metal conduit.
5) Wire and conduit size varies per insulation and termination listing.
6) Amperages listed are permitted for system voltage ranges of 220 to 240 and 440 to 480.
7) Calculations apply only to induction type, squirrel-cage, and wound-rotor motors.

THREE-PHASE AC MOTOR REQUIREMENTS *(cont.)*

Motor HP	Rated Volts	FLC Amps	OCPD Size	Starter Size	Heater Amps	Wire Size	Conduit Size
20	230	54.0	100	3	62.10	4	1"
	460	27.0	50	2	31.05	10	¾"
25	230	68.0	100	3	78.20	4	1½"
	460	34.0	50	2	39.10	8	1"
30	230	80.0	125	3	92.00	3	1½"
	460	40.0	70	3	46.00	8	1"
40	230	104.0	175	4	119.60	1	1½"
	460	52.0	100	3	59.80	6	1"
50	230	130.0	200	4	149.50	2/0	2"
	460	65.0	150	3	74.75	4	1½"
60	230	154.0	250	5	177.10	3/0	2"
	460	77.0	200	4	88.55	3	1½"
75	230	192.0	300	5	220.80	250	2½"
	460	96.0	200	4	110.40	1	1½"
100	230	248.0	400	5	285.20	350	3"
	460	124.0	200	4	142.60	1/0	2"
125	230	312.0	500	6	358.80	600	3½"
	460	156.0	250	5	179.40	3/0	2"
150	230	360.0	600	6	414.00	700	4"
	460	180.0	300	5	207.00	4/0	2½"

Notes:

1) Overcurrent device may be increased due to starting current and load conditions.
2) Wire size based on 75°C.
3) Overload heater based on motor nameplate.
4) Conduit size based on rigid metal conduit.
5) Wire and conduit size varies per insulation and termination listing.
6) Amperages listed are permitted for system voltage ranges of 220 to 240 and 440 to 480.
7) Calculations apply only to induction type, squirrel-cage, and wound-rotor motors.

FULL-LOAD CURRENTS IN AMPS FOR SINGLE-PHASE AC MOTORS

HP	115 V	200 V	208 V	230 V
1/6	4.4	2.5	2.4	2.2
1/4	5.8	3.3	3.2	2.9
1/3	7.2	4.1	4.0	3.6
1/2	9.8	5.6	5.4	4.9
3/4	13.8	7.9	7.6	6.9
1	16.0	9.2	8.8	8.0
1 1/2	20.0	11.5	11.0	10.0
2	24.0	13.8	13.2	12.0
3	34.0	19.6	18.7	17.0
5	56.0	32.2	30.8	28.0
7 1/2	80.0	46.0	44.0	40.0
10	100.0	57.5	55.0	50.0

Note: Amperages listed are permitted for system voltage ranges of 110–120 and 220–240.

SMALL MOTOR GUIDE

AC, 115 V, 60 hz, Single-Phase

Motor HP	FLC in Amps		RPM Speed
	115 V	230 V	
1/20	2.50	1.25	1,550
1/15	2.80	1.40	1,550
1/12	3.20	1.60	850
	4.10	2.05	1,550
	2.80	1.40	1,725
1/10	4.00	2.00	1,050
	3.50	1.75	1,550
1/8	3.80	1.90	1,140
	2.50	1.25	1,725

FULL-LOAD CURRENTS IN AMPS FOR DIRECT CURRENT MOTORS

HP	90 V	120 V	180 V	240 V	500 V	550 V
¼	4.0	3.1	2.0	1.6	—	—
⅓	5.2	4.1	2.6	2.0	—	—
½	6.8	5.4	3.4	2.7	—	—
¾	9.6	7.6	4.8	3.8	—	—
1	12.2	9.5	6.1	4.7	—	—
1½	—	13.2	8.3	6.6	—	—
2	—	17.0	10.8	8.5	—	—
3	—	25.0	16.0	12.2	—	—
5	—	40.0	27.0	20.0	—	—
7½	—	58.0	—	29.0	13.6	12.2
10	—	76.0	—	38.0	18.0	16.0
15	—	—	—	55.0	27.0	24.0
20	—	—	—	72.0	34.0	31.0
25	—	—	—	89.0	43.0	38.0
30	—	—	—	106.0	51.0	46.0
40	—	—	—	140.0	67.0	61.0
50	—	—	—	173.0	83.0	75.0
60	—	—	—	206.0	99.0	90.0
75	—	—	—	255.0	123.0	111.0
100	—	—	—	341.0	164.0	148.0
125	—	—	—	425.0	205.0	185.0
150	—	—	—	506.0	246.0	222.0
200	—	—	—	675.0	330.0	294.0

Note: Amperages listed are for motors running at base speed.

DIRECT CURRENT MOTOR REQUIREMENTS

Motor HP	Full-Load Current in amps		Minimum Conduit Size in inches		Minimum Wire Size AWG/Kcmil	
	115 V	230 V	115 V	230 V	115 V	230 V
1	8.4	4.2	½	½	14	14
1½	12.5	6.3	½	½	12	14
2	16.1	8.3	¾	½	10	14
3	23	12.3	¾	½	8	12
5	40	19.8	1	¾	6	10
7½	58	28.7	1¼	1	3	6
10	75	38	1½	1	1	6
15	112	56	2	1¼	2/0	4
20	140	74	2	1½	3/0	1
25	184	92	2½	2	300	1/0
30	220	110	3	2	400	2/0
40	292	146	3½	2½	700	4/0
50	360	180	4	2½	1,000	300
60	–	215	–	3	–	400
75	–	268	–	3½	–	600
100	–	355	–	4	–	1,000

DC MOTOR PERFORMANCE CHARACTERISTICS

Performance Characteristics	Voltage 10% below Rated Voltage		Voltage 10% above Rated Voltage	
	Shunt	Compound	Shunt	Compound
Starting Torque	−15%	−15%	+15%	+15%
Speed	−5%	−6%	+5%	+6%
Current	+12%	+12%	−8%	−8%
Field Temperature	Increases	Decreases	Increases	Increases
Armature Temperature	Increases	Increases	Decreases	Decreases
Commutator Temperature	Increases	Increases	Decreases	Decreases

MAXIMUM ACCELERATION TIME

Motor Frame Number	Maximum Acceleration Time (in seconds)
48 and 56	8
143–286	10
324–326	12
364–505	15

HORSEPOWER RATINGS FOR 240 VOLT AC SAFETY SWITCHES

General Duty 240 V Applications

Rating in Amps	Fusible				Nonfusible	
	Standard		Maximum		Maximum	
	1φ	3φ	1φ	3φ	1φ	3φ
30	1½	3	3	7½	3	7½
60	3	7½	10	15	10	15
100	7½	15	15	30	15	30
200	15	25	—	60	—	60
400	—	50	—	125	—	125
600	—	75	—	200	—	150

Heavy Duty 240 V Applications

Rating in Amps	Fusible				Nonfusible	
	Standard		Maximum		Maximum	
	1φ	3φ	1φ	3φ	1φ	3φ
30	1½	3	3	7½	5	10
60	3	7½	10	15	10	20
100	7½	15	15	30	20	40
200	15	25	15	60	—	60
400	—	50	—	125	—	125
600	—	75	—	200	—	200
800	50	—	50	250	—	250
1,200	50	—	50	—	—	250

Notes: Horsepower ratings vary by manufacturer. Check the manual or the disconnect itself for the actual rating.

The maximum horsepower rating applies to time-delay fuses and the standard to non-time-delay fuses.

HORSEPOWER RATINGS FOR 240 V
AC SAFETY SWITCHES *(cont.)*

Heavy Duty 480 V Applications

| Rating in Amps | Fusible | | Nonfusible |
| | Standard | Maximum | Maximum |
	3ϕ	3ϕ	3ϕ
30	5	15	20
60	15	30	50
100	25	60	75
200	50	125	125
400	100	250	250
600	150	400	400
800	200	500	500
1,200	200	500	500

Notes: Horsepower ratings vary by manufacturer. Check the manual or the disconnect itself for the actual rating.

The maximum horsepower rating applies to time-delay fuses and the standard to non-time-delay fuses.

Additional Horsepower Rating Notes:

1) For a safety switch to be horsepower rated, it must be able to interrupt a motor's locked-rotor current and capable of handling six times the motor's full-load current rating.
2) The horsepower rating of IEC motor starters must be handled differently in that they do not have the same reserve capacity as NEMA® starters.
3) Derate the horsepower if the motor is to perform jogging or plugging duty.
4) When IEC motors are used and are rated in watts, use the following guidelines for converting to equivalent horsepower.
   720 watts = 1 horsepower or
   1 kilowatt = 1.34 horsepower
5) Remember that these figures will vary based on motor efficiency.

MAXIMUM RATING OR SETTING OF MOTOR BRANCH CIRCUIT OVERCURRENT PROTECTIVE DEVICES IN PERCENT OF FULL-LOAD CURRENT[1]

Type of Motor	Non-Time-Delay Fuse[2]	Dual Element (Time-Delay) Fuse[2]	Instantaneous Trip Breaker	Inverse Time Breaker[3]
Single-phase	300%	175%	800%	250%
AC polyphase except wound-rotor	300%	175%	800%	250%
Squirrel cage except Design B (energy-efficient)	300%	175%	800%	250%
Design B (energy-efficient)	300%	175%	1100%	250%
Synchronous[4]	300%	175%	800%	250%
Wound-rotor	150%	150%	800%	150%
Direct current (constant voltage)	150%	150%	250%	150%

[1]For certain exceptions to the values specified, refer to the NEC®.

[2]The values in the Non-Time-Delay Fuse column apply to Time-Delay Class CC fuses.

[3]The values given in the last column also cover the ratings of nonadjustable inverse time types of circuit breakers that may be modified (refer to the NEC® for additional information).

[4]Synchronous motors of the low-torque, low-speed type (450 rpm or lower) that start unloaded do not require a fuse rating or circuit-breaker setting in excess of 200 percent of full-load current.

OVERLOAD DEVICE PLACEMENT

Type of Motor	Motor Supply Circuit Type	Number and Placement in Circuit for Overload Devices Such as Trip Coils or Relays
Single-phase AC or DC	2-wire, single-phase AC or DC, ungrounded	1 in either conductor
Single-phase AC or DC	2-wire, single-phase AC or DC, one conductor grounded	1 in ungrounded conductor
Single-phase AC or DC	3-wire, single-phase AC or DC, grounded neutral conductor	1 in either ungrounded conductor
Single-phase AC	Any three-phase	1 in ungrounded conductor
Two-phase AC	3-wire, two-phase AC, ungrounded	2, one in each phase
Two-phase AC	3-wire, two-phase AC, one conductor grounded	2 in ungrounded conductors
Two-phase AC	4-wire, two-phase AC, grounded or ungrounded	2, one per phase in ungrounded conductors
Two-phase AC	Grounded neutral or 5-wire, two-phase AC, ungrounded	2, one per phase in any ungrounded phase wire
Three-phase AC	Any three-phase	3, one in each phase*

*Exception: Not required where overload protection is provided by other means.

6-15

HORSEPOWER FORMULAS

Current and Voltage Known

$$HP = \frac{E \times I \times E_{FF}}{746}$$

where
HP = horsepower
I = current (amps)
E = voltage (volts)
E_{FF} = efficiency
746 = constant

Speed and Torque Known

$$HP = \frac{RPM \times T}{5252}$$

where
HP = horsepower
RPM = revolutions per minute
T = torque (ft-lb)
5252 = constant

EFFICIENCY FORMULAS

Input and Output Power Known

$$E_{FF} = \frac{P_{out}}{P_{in}}$$

where
E_{FF} = efficiency (%)
P_{out} = output power (watts)
P_{in} = input power (watts)

Horsepower and Power Loss Known

$$E_{FF} = \frac{746 \times HP}{(746 \times HP) + W_l}$$

where
E_{FF} = efficiency (%)
746 = constant
HP = horsepower
W_l = watts lost

VOLTAGE UNBALANCE

$$V_u = \frac{V_d}{V_a} \times 100$$

where
V_u = voltage unbalance (%)
V_d = voltage deviation (volts)
V_a = voltage average (volts)
100 = constant

TEMPERATURE CONVERSIONS

Convert °C to °F

$$°F = (1.8 \times °C) + 32$$

Convert °F to °C

$$°C = \frac{(°F - 32)}{1.8}$$

LOCKED ROTOR CURRENT FORMULAS

Apparent, 1φ	Apparent, 3φ	True, 1φ	True, 3φ
$LRC = \dfrac{1,000 \times HP \times KVA/HP}{V}$	$LRC = \dfrac{1,000 \times HP \times KVA/HP}{V \times \sqrt{3}}$	$LRC = \dfrac{1,000 \times HP \times KVA/HP}{V \times PF \times E_{FF}}$	$LRC = \dfrac{1,000 \times HP \times KVA/HP}{V \times \sqrt{3} \times PF \times E_{FF}}$
where	where	where	where
LRC = locked rotor current (in amps)	LRC = locked rotor current (in amps)	LRC = locked rotor current (in amps)	LRC = locked rotor current (in amps)
1,000 = multiplier for kilo	1,000 = multiplier for kilo	1,000 = multiplier for kilo	1,000 = multiplier for kilo
HP = horsepower	HP = horsepower	HP = horsepower	HP = horsepower
KVA/HP = kilovolt amps per horsepower	KVA/HP = kilovolt amps per horsepower	KVA/HP = kilovolt amps per horsepower	KVA/HP = kilovolt amps per horsepower
V = volts	V = volts	V = volts	V = volts
	$\sqrt{3} = 1.732$	PF = power factor	$\sqrt{3} = 1.732$
		E_{FF} = motor efficiency	PF = power factor
			E_{FF} = motor efficiency

GEAR REDUCER FORMULAS

Output Torque	Output Speed	Output Horsepower
$O_T = I_T \times R_R \times R_E$ where O_T = output torque (ft-lb) I_T = input torque (ft-lb) R_R = gear reducer ratio R_E = reducer efficiency (%)	$O_s = \dfrac{I_S}{R_R} \times R_E$ where O_s = output speed (rpm) I_S = input speed (rpm) R_R = gear reducer ratio R_E = reducer efficiency (%)	$O_{HP} = I_{HP} \times R_E$ where O_{HP} = output horsepower I_{HP} = input horsepower R_E = reducer efficiency (%)

MOTOR TORQUE FORMULAS

Torque	Starting Torque	Nominal Torque Rating
$T = \dfrac{HP \times 5{,}252}{RPM}$ where T = torque (ft-lb) HP = horsepower $5{,}252$ = constant RPM = revolutions per minute	$T = \dfrac{HP \times 5{,}252 \times E_{EF}}{RPM}$ where T = torque (ft-lb) HP = horsepower $5{,}252$ = constant RPM = revolutions per minute E_{FF} = motor efficiency (%)	$T = \dfrac{HP \times 63{,}025}{RPM}$ where T = nominal torque rating (in-lb) HP = horsepower $63{,}025$ = constant RPM = revolutions per minute

MOTOR TORQUE PER MOTOR SPEED

Motor HP	RPM Speeds of Motors			
	68	100	155	190
1	927	630	407	332
1½	1,390	945	610	498
2	1,854	1,261	813	663
3	2,781	1,891	1,220	995
5	4,634	3,151	2,033	1,659
7½	6,951	4,727	3,050	2,488
10	9,268	6,303	4,066	3,317
15	13,903	9,454	6,099	4,976
20	18,537	12,605	8,132	6,634
25	23,171	15,756	10,165	8,293
30	27,805	18,908	12,198	9,951
40	37,074	25,210	16,265	13,268
50	46,342	31,513	20,331	16,586
60	55,610	37,815	24,397	19,903
70	64,879	44,118	28,463	23,220
80	74,147	50,420	32,529	26,537
90	83,415	56,723	36,595	29,854
100	92,684	63,025	40,661	33,171
125	115,855	78,781	50,827	41,464
150	139,026	94,538	60,992	49,757
175	162,197	110,294	71,157	58,049
200	185,368	126,050	81,323	66,342
225	208,539	141,806	91,488	74,635
250	231,710	157,563	101,653	82,928
275	254,881	173,319	111,819	91,220
300	278,051	189,075	121,984	99,513
350	324,393	220,588	142,315	116,099
400	370,735	252,100	162,645	132,684
450	417,077	283,613	182,976	149,270
500	463,419	315,125	203,306	165,855
550	509,761	346,638	223,637	182,441
600	556,103	378,150	243,968	199,026

The torque values above are measured as inch-pounds of force. To obtain torque in foot-pounds of force, divide these values by 12.

MOTOR TORQUE PER MOTOR SPEED (cont.)

Motor HP	RPM Speeds of Motors			
	500	750	850	1,000
1	126	84	74	63
1½	189	126	111	95
2	252	168	148	126
3	378	252	222	189
5	630	420	371	315
7½	945	630	556	473
10	1,261	840	741	630
15	1,891	1,261	1,112	945
20	2,521	1,691	1,483	1,261
25	3,151	2,101	1,854	1,576
30	3,782	2,521	2,224	1,891
40	5,042	3,361	2,966	2,521
50	6,303	4,202	3,707	3,151
60	7,563	5,042	4,449	3,782
70	8,824	5,882	5,190	4,412
80	10,084	6,723	5,932	5,042
90	11,345	7,563	6,673	5,672
100	12,605	8,403	7,415	6,303
125	15,756	10,504	9,268	7,878
150	18,908	12,605	11,122	9,454
175	22,059	14,706	12,976	11,029
200	25,210	16,807	14,829	12,605
225	28,361	18,908	16,683	14,181
250	31,513	21,008	18,537	15,756
275	34,664	23,109	20,390	17,332
300	37,815	25,210	22,244	18,908
350	44,118	29,412	25,951	22,059
400	50,420	33,613	29,659	25,210
450	56,723	37,815	33,366	28,361
500	63,025	42,017	37,074	31,513
550	69,328	46,218	40,781	34,664
600	75,630	50,420	44,488	37,815

The torque values above are measured as inch-pounds of force. To obtain torque in foot-pounds of force, divide these values by 12.

MOTOR TORQUE PER MOTOR SPEED (cont.)

Motor HP	RPM Speeds of Motors			
	1,050	1,550	1,725	3,450
1	60	41	37	18
1½	90	61	55	27
2	120	81	73	37
3	180	122	110	55
5	300	203	183	91
7½	450	305	274	137
10	600	407	365	183
15	900	610	548	274
20	1,200	813	731	365
25	1,501	1,017	913	457
30	1,801	1,220	1,096	548
40	2,401	1,626	1,461	731
50	3,001	2,033	1,827	913
60	3,601	2,440	2,192	1,096
70	4,202	2,846	2,558	1,279
80	4,802	3,253	2,923	1,461
90	5,402	3,660	3,288	1,644
100	6,002	4,066	3,654	1,827
125	7,503	5,083	4,567	2,284
150	9,004	6,099	5,480	2,740
175	10,504	7,116	6,394	3,197
200	12,005	8,132	7,307	3,654
225	13,505	9,149	8,221	4,110
250	15,006	10,165	9,134	4,567
275	16,507	11,182	10,047	5,024
300	18,007	12,198	10,961	5,480
350	21,008	14,231	12,788	6,394
400	24,010	16,265	14,614	7,307
450	27,011	18,298	16,441	8,221
500	30,012	20,331	18,268	9,134
550	33,013	22,364	20,095	10,047
600	36,014	24,397	21,922	10,961

The torque values above are measured as inch-pounds of force. To obtain torque in foot-pounds of force, divide these values by 12.

TIPS ON SELECTING MOTORS

First step in selecting a motor for a particular drive is to obtain data listed below. Motors generally operate at best power factor and efficiency when fully loaded.

TORQUE: Starting torque needed by load must be less than required starting torque of proposed motor. Motor torque must never fall below driven machine's torque needs in going from standstill to full speed.

Torque requirements of some loads may fluctuate between wide limits. Although average torque may be low, many torque peaks may be well above full-load torque. If load torque impulses are repeated frequently (air compressor), it's best to use a high-slip motor with a flywheel. But if load is generally steady at full load, you can use a more efficient low-slip motor. Only in this case any intermittent load peaks are taken directly by the motor and reflect back into power system. Also breakdown (maximum motor torque) must be higher than load-peak torque.

ENCLOSURES: Atmospheric conditions surrounding motor determine enclosure used. The more enclosed a motor, the more it costs and the hotter it tends to run. Totally-enclosed motors may require a larger frame size for a given hp than open or protected motors.

INSULATION: This, likewise, is determined by surrounding atmosphere and operating temperature. Ambient (room) temperature is generally assumed to be 40°C. Total temperature that the motor reaches directly influences insulation life. Each 10°C rise in max temp halves effective life of *Class A* and *B* insulations.

Motor temperature rise is maximum temperature (over ambient) measured with an external thermometer. "Hot-spot" allowance takes care of temperature difference between external reading and the hottest spot within windings. Service-factor allows for continuous overload—15% for general-purpose open, protected, or drip-proof motors.

VARIABLE CYCLE: Where load varies according to some regular cycle, it would not be economical to select a motor that matches the peak load.

Instead (for AC induction motors where speed is not varied) calculate hp needed on the *root-means-square* (rms) basis. *Rms* hp is equivalent to the continuous hp that would produce the same heat in motor as cycle operation. Torque-speed relation of motor should still match that of load.

FACTS TO CONSIDER

REQUIREMENTS OF DRIVEN MACHINE:
1. HP needed
2. Torque range
3. Operating cycle: frequency of starts and stops
4. Speed
5. Operating position: horizontal, vertical, or tilted
6. Direction of rotation
7. Endplay and thrust
8. Ambient (room) temperature
9. Surrounding conditions: water, gas, corrosion, dust, outdoor, etc.

ELECTRICAL SUPPLY:
1. Voltage of power system
2. Number of phases
3. Frequency
4. Limitations on starting current
5. Effect of demand, energy on power rates

DIRECT CURRENT MOTOR DATA

Chief reason for using DC motors, assuming normal power source is AC, lies in the wide and economical ranges possible of speed control and starting torques. But for constant-speed service, AC motors are generally preferred because they are more rugged and have lower first cost.

STARTING TORQUE: With a shunt motor, torque is proportional to armature current, because field flux remains practically constant for a given setting of the field rheostat. However, the flux of a series field is affected by the current through it. At light loads, flux varies directly with a current, so torque varies as the square of the current. The compound motor (usually cumulative) lies in between the shunt and series motors as to torque.

Upper limit of current input on starting is usually 1.5 to 2 times full-load current to avoid overheating the commutator, excessive fedder drops, or peaking generator. Shunt-motor starting boxes usually allow 125% current at first notch. So motor can develop 125% starting torque. Series motors can develop higher starting torques at same current, since torque increases as current squared. Compound motors develop starting torques higher than shunt motors according to amount of compounding.

SPEED CONTROL: Shunt motor speeds drop only slightly (5% or less) from no load to full load. Decreasing field current raises speed; increasing field reduces speed. But speed is still practically constant for any one field setting. Speed can be controlled by resistance in the armature circuit but regulation is poor.

Series motor speeds decrease much more with increased load, and, conversely, begin to race at low loads, dangerously so if load is completely removed. Speed can be reduced by adding resistance into the armature circuit, increased by shunting the series field with resistance or short-circuiting series turns.

Compound motors have less constant speed than shunt motors and can be controlled by shunt-field rheostat.

APPLICATIONS FOR DIRECT CURRENT AND SINGLE-PHASE MOTORS

Speed Regulation	Speed Control	Starting Torque	Pull-out Torque	Motor Applications
Series				
Varies inversely as the load. Races on light loads and full voltage	Zero to maximum, depending on control and load	High. Varies as square of the voltage. Limited by commutation, heating, and line capacity	High. Limited by commutation, heating, and line capacity	Where high starting torque is required and speed can be regulated. Traction, bridges, hoists, gates, car dumpers, car retarders, etc.
Shunt				
Drops 3 to 5% from no load to full load	Any desired range depending on motor design and type of system	Good. With constant field, varies directly as voltage applied to armature	High. Limited by commutation, heating, and line capacity	Where constant or adjustable speed is required and starting conditions are not severe. Fan, blowers, centrifugal pumps, conveyers, wood-working machines, metal-working machines, elevators
Compound				
Drops 7 to 20% from no load to full load depending on amount of compounding	Any desired range, depending on motor design and type of control	Higher than for shunt, depending on amount of compounding	High. Limited by commutation, heating, and line capacity	Where high starting torque combined with fairly constant speed is required. Plunger pumps, punch presses, shears, bending rolls, geared elevators, conveyers, hoists

APPLICATIONS FOR DIRECT CURRENT AND SINGLE-PHASE MOTORS *(cont.)*

Speed Regulation	Speed Control	Starting Torque	Pull-out Torque	Motor Applications
CAPACITOR				
Drops 5% for large to 10% for small sizes	None	150 to 350% of full load, depending upon design and size	150% for large to 200% for small sizes	Constant speed service for any starting duty, and quiet operation, where polyphase current cannot be used
COMMUTATOR-TYPE				
Drops 5% for large to 10% for small sizes	Repulsion-induction, none. Brush shifting types 4 to 1 at full load	250% for large to 350% for small sizes	150% for large to 250% for small sizes	Constant speed service for any starting duty, where speed control is required and polyphase current cannot be used
SPLIT-PHASE				
Drops about 10% from no load to full load	None	75% for large to 175% for small sizes	150% for large to 200% for small sizes	Constant speed service where starting is easy. Small fans, centrifugal pumps, and light running machines, where polyphase current is not available

SQUIRREL-CAGE MOTOR DATA

Squirrel-cage induction motors are classed by National Electrical Manufacturers Association (NEMA) according to locked-rotor torque, breakdown torque, slip, starting current, etc. Common types are Class B, C, and D.

CLASS B is the most common type, has normal starting torque, low starting current. Locked-rotor torque (minimum torque at standstill and full voltage) is not less than 100% full-load for 2- and 4-pole motors, 200 hp and less; 40 to 75% for larger 2-pole motors; 50 to 125% for larger 4-pole motors.

CLASS C features high starting torque (locked-rotor over 200%), low starting current. Breakdown torque not less than 190% full-load torque. Slip at full load is between 1½ and 3%.

CLASS D have high slip, high starting torque, low starting current; are used on loads with high intermittent peaks. Driven machine usually has high-inertia flywheel. At no load motor has little slip; when peak load is applied, motor slip increases. Speed reduction lets driven machine absorb energy from flywheel rather than power line.

STARTING Full-voltage, across-the-line starting is used where power supply permits and full-voltage torque and acceleration are not objectionable. Reduction in starting kVA cuts locked-rotor and accelerating torques.

APPLICATIONS FOR SQUIRREL-CAGE MOTORS

Speed Regulation	Speed Control	Starting Torque	Pull-out Torque	Motor Applications
GENERAL-PURPOSE SQUIRREL-CAGE (Class B)				
Drops about 3% for large to 5% for small sizes	None, except multi-speed types designed for 2 to 4 fixed speeds	200% of full load for 2-pole to 105% for 16-pole designs	200% of full load	Constant-speed service where starting torque is not excessive. Fans, blowers, rotary compressors, centrifugal pumps
HIGH-TORQUE SQUIRREL-CAGE (Class C)				
Drops about 3% for large to 6% for small sizes	None, except multi-speed types designed for 2 to 4 fixed speeds	250% of full load for high-speed to 200% for low-speed designs	200% of full load	Constant-speed service where fairly high starting torque is required at infrequent intervals with starting current of about 400% of full load. Reciprocating pumps and compressors, crushers, etc.
HIGH-SLIP SQUIRREL-CAGE (Class D)				
Drops about 10 to 15% from no load to full load	None, except multi-speed types designed for 2 to 4 fixed speeds	225 to 300% of full load, depending on speed with rotor resistance	200%. Will usually not stall until loaded to max torque, which occurs at standstill	Constant-speed service and high starting torque, if starting is not too frequent, and for taking high peak loads with or without flywheels. Punch presses, shears, elevators, etc.

WOUND-ROTOR MOTOR DATA

The wound-rotor (slip-ring) induction motor's rotor winding connects through slip-rings to an external resistance that is cut in and out by a controller.

RESISTANCE vs. TORQUE: Resistance of rotor winding affects torque developed at any speed. A high-resistance rotor gives high starting torque with low starting current. But low slip at full load, good efficiency, and moderate rotor heating takes a low-resistance rotor. Left-hand curves show rotor-resistance effect on torque. With all resistance in, R_1, full-load starting torque is developed at less than 150% full-load current. Successively shorting out steps, at standstill, develops about 225% full-load torque at R_4. Cutting out more reduces standstill torque. Motor operates like a squirrel-cage motor when all resistance is shorted out.

SPEED CONTROL: Having resistance left in decreases speed regulation. Right-hand curves for a typical motor show that with only two steps shorted out, motor operates at 65% synchronous speed because motor torque equals load torque at that speed. But if load torque drops to 50%, motor shoots forward to about 65% synchronous speed.

However, slip rings are normally shorted after motor comes up to speed. Or, for short-time peak loads, motor is operated with a step or two of resistance cut in. At light loads, motor runs near synchronous speed. When peak loads come on, speed drops; flywheel effect of motor and load cushions power supply from load peak.

OTHER FEATURES: In addition to high-starting-torque, low-starting-current applications, wound-rotor motors are used (1) for high-inertia loads where high slip losses that would have to be dissipated in the rotor of a squirrel-cage motor, in coming up to speed, can be given off as heat in wound-rotor's external resistance (2) where frequent starting, stopping, and speed control are needed (3) for continuous operation at reduced speed (e.g., boiler draft fan — combustion control varies external resistance to adjust speed, damper regulates air flow between step speeds and below 50% speed).

CONTROLS: Across-the-line starters are used with proper protection (fuses, breakers, etc.) and secondary control with 5 to 7 resistance steps.

SYNCHRONOUS MOTOR DATA

Synchronous motors run at a fixed or synchronous speed determined by line frequency and number of poles in machine. (Rpm = 120 x frequency/number or poles.) Speed is kept constant by locking action of an externally excited DC field. Efficiency is 1 to 3% higher than that of same-size-and-speed induction or DC motors. Also synchronous motors can be operated at power factors from 1.0 down to 0.2 leading for plant power–factor correction. Standard ratings are 1.0 and 0.9 leading PF; machines rated down near 0.2 leading are called synchronous condensers.

STARTING: Pure synchronous motors are not self-starting; so in practice they are built with damper or amortisseur windings. With the field coil shorted through discharge resistor, damper winding acts like a squirrel-cage rotor to bring motor practically to synchronous speed; then field is applied and motor pulls into synchronism, providing motor has developed sufficient "pull-in" torque. Once in synchronism, motor keeps constant speed as long as load torque does not exceed maximum or "pull-out" torque; then machine drops out of synchronism. Driven machine is usually started without load. Low-speed motors may be direct connected.

FIELD AND PF: While motors are rated for specific power factors . . . at constant power, increasing DC field current causes power factor to lead, decreasing field tends to make PF lag. But either case increases copper losses.

TYPES: Polyphase synchronous motors in general use are: (1) high-speed motors, 500 rpm up, (a) general-purpose, 500 to 1,800 rpm, 200 hp and below, (b) high-speed units over 200 hp including most 2-pole motors; (2) low-speed motors below 500 rpm; (3) special high-torque motors.

COSTS: Small high-speed synchronous motors cost more than comparable induction motors, but with low-speed and large high-speed motor, costs favor the synchronous motor. Cost of leading-PF motors increases approximately inversely proportional to the decrease from unity power factor.

APPLICATIONS FOR WOUND-ROTOR AND SYNCHRONOUS MOTORS

Speed Regulation	Speed Control	Starting Torque	Pull-out Torque	Motor Applications
WOUND-ROTOR				
With rotor rings short-circuited, drops about 3% for large to 5% for small sizes	Speed can be reduced to 50% by rotor resistance to obtain stable operation. Speed varies inversely as load	Up to 300%, depending on external resistance in rotor circuit and how distributed	200%, when rotor slip rings are short-circuited	Where high starting torque with low starting current or where limited speed control is required. Fans, centrifugal and plunger pumps, compressors, conveyers, hoists, cranes
SYNCHRONOUS				
Constant	None, except special motors designed for 2 fixed speeds	40% for slow to 160% for medium speed, 80% PF designs. Special designs develop higher torques	Unity of motors 170%; 80% PF motors 225%. Special designs up to 300%	For constant-speed service, direct connection to slow speed machines, and where power factor correction is required

TYPES OF MOTOR ENCLOSURES

OPEN-TYPE
has full openings in frame and endbells for maximum ventilation, is lowest-cost enclosure.

SEMI-PROTECTED
has screens in top openings to keep out falling objects. PROTECTED has screens in bottom too.

DRIP-PROOF
has upper parts covered to keep out drippings falling at angle not over 15° from vertical.

SPLASH-PROOF
is baffled at bottom to keep out particles coming at angle not over 100° from vertical.

TOTALLY-ENCLOSED
can be non-ventilated, separately ventilated, or explosion proof for hazardous atmospheres.

FAN-COOLED
has a totally-enclosed motor with double covers. Fan, behind vented outer shroud, is run by motor.

TYPES OF INSULATION

CLASS A (cotton, silk, paper, or other organics impregnated with insulating varnish) is considered standard for most applications, allows 105° C total temperature.

40°C ambient
40°C rise by thermometer
15°C "hot-spot" allowance
10°C service factor
_____
105°C total temperature

CLASS B (mica, asbestos, fiber-glass, other inorganics) allows 130°C total temperature.

40°C ambient
70°C rise by thermometer
20°C "hot-spot" allowance
_____
130°C total temperature

CLASS H (including silicone family) is for special high-temperature applications.

SPECIAL CLASS A is highly resistant, but not "proof," against severe moisture, dampness; conductive, corrosive, or abrasive dusts and vapors.

TROPICAL is for excessive moisture, high ambients, corrosion, fungus, vermin, insects.

MOTOR FRAME LETTERS

Letter	Designation
G	Gasoline pump motor
K	Sump pump motor
M and N	Oil burner motor
S	Standard short shaft for direct connection
T	Standard dimensions established
U	Previously used as frame designation
Y	Special mounting dimensions required
Z	Shaft extension

SHAFT COUPLING SELECTIONS

Coupling Number	Rated Torque (in-lb)	Maximum Shock Torque (in-lb)
10-101-A	16	45
10-102-A	36	100
10-103-A	80	220
10-104-A	132	360
10-105-A	176	480
10-106-A	240	660
10-107-A	325	900
10-108-A	525	1,450
10-109-A	875	2,450
10-110-A	1,250	3,500
10-111-A	1,800	5,040
10-112-A	2,200	6,160

MOTOR V-BELTS

No. 0 Section
"2L"

No. 1 Section
"3L"

No. 2 Section
"4L"
A

No. 3 Section
"5L"
B

V-BELT PER MOTOR SIZE

3/8"
9.5 mm

Up to 0.76 kW
1 HP

1/2"
11.7 mm

0.56 to 4 kW
5 HP

21/32"
16.7 mm

2.5 kW and up
3 HP

NEMA MOTOR FRAME DIMENSIONS

Frame No.	All Dimension in Inches						
	D	E	F	U	V	M+N	Keyway
42	2⅝	1¾	²⁷⁄₃₂	⅜	—	4¹⁄₃₂	—
48	2	2⅛	1⅜	½	—	5⅜	—
56	3½	2⁷⁄₁₆	1½	⅝	—	6⅛	³⁄₁₆ × ³⁄₃₂
66	4⅛	2¹⁵⁄₁₆	2½	¾	—	7⅛	³⁄₁₆ × ³⁄₃₂
143T	3½	2¾	2	⅞	2	6½	³⁄₁₆ × ³⁄₃₂
145T	3½	2¾	2½	⅞	2	7	³⁄₁₆ × ³⁄₃₂
182	4½	3¾	2¼	⅞	2	7¼	³⁄₁₆ × ³⁄₃₂
182T	4½	3¾	2¼	1⅛	2½	7¾	¼ × ⅛
184	4½	3¾	2¾	⅞	2	7¾	³⁄₁₆ × ³⁄₃₂
184T	4½	3¾	2¾	1⅛	2½	8¼	¼ × ⅛
213	5¼	4¼	2¾	1⅛	2¾	9¼	¼ × ⅛
213T	5¼	4¼	2¾	1⅜	3⅛	9⅝	⁵⁄₁₆ × ⁵⁄₃₂
215	5¼	4¼	3½	1⅛	2¾	10	¼ × ⅛
215T	5¼	4¼	3½	1⅜	3⅛	10⅜	⁵⁄₁₆ × ⁵⁄₃₂
254T	6¼	5	4⅛	1⅝	3¾	12⅝	⅜ × ³⁄₁₆
254U	6¼	5	4⅛	1⅜	3½	12⅛	⁵⁄₁₆ × ⁵⁄₃₂
256T	6¼	5	5	1⅝	3¾	13¼	⅜ × ³⁄₁₆
256U	6¼	5	5	1⅜	3½	13	⁵⁄₁₆ × ⁵⁄₃₂
284T	7	5½	4¾	1⅞	4⅜	14⅛	½ × ¼
284TS	7	5½	4¾	1⅝	3	12¾	⅜ × ³⁄₁₆
284U	7	5½	4¾	1⅝	4⅝	14⅜	⅜ × ³⁄₁₆
286T	7	5½	5½	1⅞	4⅜	14⅞	½ × ¼
286U	7	5½	5½	1⅝	4⅝	15⅛	⅜ × ³⁄₁₆
324T	8	6¼	5¼	2⅛	5	15¾	½ × ¼
324U	8	6¼	5¼	1⅞	5⅝	16⅛	½ × ¼
326T	8	6¼	6	2⅛	5	16½	½ × ¼
326TS	8	6¼	6	1⅞	3½	15	½ × ¼
326U	8	6¼	6	1⅞	5⅝	16⅞	½ × ¼
364T	9	7	5⅝	2⅜	5⅝	17⅜	⅝ × ⁵⁄₁₆
364U	9	7	5⅝	2⅛	6⅛	17⅞	½ × ¼
365T	9	7	6⅛	2⅜	5⅝	17⅞	⅝ × ⁵⁄₁₆
365U	9	7	6⅛	2⅛	6⅛	18⅜	½ × ¼
404T	10	8	6⅛	2⅞	7	20	¾ × ⅜
404U	10	8	6⅛	2⅜	6⅞	19⅞	⅝ × ⁵⁄₁₆
405T	10	8	6⅞	2⅞	7	20¾	¾ × ⅜
405U	10	8	6⅞	2⅜	6⅞	20⅝	⅝ × ⁵⁄₁₆
444T	11	9	7¼	3⅜	8¼	23¼	⅞ × ⁷⁄₁₆
444U	11	9	7¼	2⅞	8⅜	23⅜	¾ × ⅜
445T	11	9	8¼	3⅜	8¼	24¼	⅞ × ⁷⁄₁₆
445U	11	9	8¼	2⅞	8⅜	24⅜	¾ × ⅜

Standards established by National Electrical Manufacturers Association.

FRONTAL VIEW OF A TYPICAL MOTOR

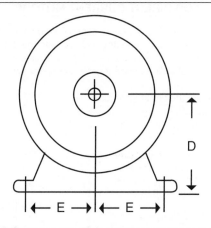

REFERENCE PAGE 6–34 FOR DIMENSIONS

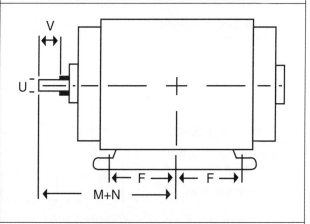

SIDE VIEW OF A TYPICAL MOTOR

ROTATION AND TERMINAL MARKINGS
FOR DIRECT CURRENT MOTORS

Terminal markings identify connections from outside circuits. Facing the commutator end (the end opposite the drive), the direction of the shaft rotation is counterclockwise. The standard rotation of a direct current generator is clockwise.

A_1 and A_2 are the armature leads.
S_1 and S_2 are the series-field leads.
F_1 and F_2 are the shunt-field leads.

FOR SERIES- WOUND MOTORS:
To change rotation, reverse either armature leads or series leads. Do not reverse both.

FOR SHUNT- WOUND MOTORS:
To change rotation, reverse either armature leads or shunt leads. Do not reverse both.

FOR COMPOUND- WOUND MOTORS:
To change rotation, reverse either armature leads or both the series and shunt leads. Do not reverse all three.

ROTATION AND TERMINAL MARKINGS
FOR SINGLE-PHASE MOTORS

There are two main groups of single-phase motors, each of which is made up of various types outlined below.

A) Split-phase:
  - Capacitor-Start
  - Resistance-Start
  - Split-Capacitor

B) Commutator:
  - Repulsion
  - Repulsion-Start
  - Series

The colors of the motor terminals are as follows:

T_1 – Blue	T_3 – Orange	T_5 – Black
T_2 – White	T_4 – Yellow	T_8 – Red

Note: To change the speed of a split-phase motor, the number of poles must be changed, but this is rarely done in the field.

SPLIT-PHASE, SQUIRREL-CAGE MOTORS

Dual Voltage

T_4 T_5 T_2 T_3 T_8 T_1

115 V
Counterclockwise rotation

T_4 T_5 T_2 T_3 T_8 T_1

230 V
Counterclockwise rotation

T_8
T_1
T_2
Main winding
Section 1
L_1
L_2
T_3
T_4
Main winding
Section 2
Starting
Winding
T_5
Centrifugal
Switch

To reverse rotation, interchange
T_5 and T_8

6-37

SPLIT-PHASE, SQUIRREL-CAGE MOTORS *(cont.)*

Resistance-Start

This motor has a resistance connected in series with the starting winding.

The centrifugal switch opens after reaching 75% of normal operating speed.

Capacitor-Start

This motor is employed where a high starting torque is required.

SPLIT-PHASE MOTOR ROTATION

Single Voltage

Clockwise
Rotation

Counterclockwise
Rotation

Dual Voltage

Low voltage

High voltage

Clockwise Rotation

Low voltage

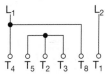

High voltage

Counterclockwise Rotation

DELTA-WOUND MOTOR CONNECTIONS 240/480 V

Low-Voltage Connection

High-Voltage Connection

WYE-WOUND MOTOR CONNECTIONS 240/480 V

Low-Voltage Connection

High-Voltage Connection

DUAL-VOLTAGE, 3φ, WYE-CONNECTED MOTORS

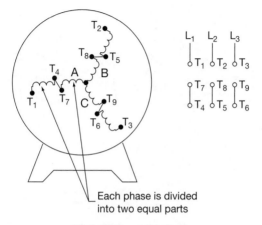

Each phase is divided
into two equal parts

High-Voltage (series)

Low-Voltage (parallel)

DUAL-VOLTAGE, 3φ, DELTA-CONNECTED MOTORS

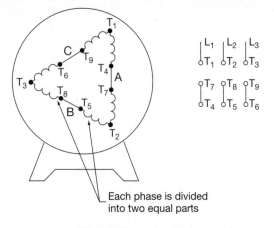

Each phase is divided into two equal parts

High-Voltage (series)

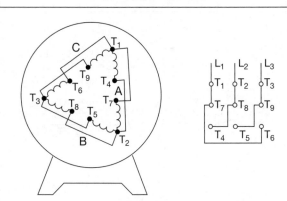

Low-Voltage (parallel)

STAR-CONNECTED, POLYPHASE MOTOR

REVERSING THREE-PHASE MOTORS

Forward   **Reverse**

Reversing direction is accomplished by interchanging any two of the three power lines. Although any two may be interchanged, the industry standard is to interchange L_1 and L_3. This is true for all 3ϕ motors including three-, six-, and nine- lead wye- and delta-connected motors.

6-43

MOTOR CONTROL CIRCUIT SCHEMATICS

Magnetic starter with one stop-start station and a pilot lamp which burns to indicate that the motor is running

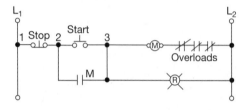

Magnetic starter with three stop-start stations

WIRING DIAGRAM OF A SINGLE-PHASE MOTOR CONNECTED TO A THREE-PHASE MOTOR STARTER

L₁ L₂ L₃

L_1 L_2 Not used L_3

Disconnect switch →

Fuse →

Start

Stop

2

Aux.

3

M

Overload heater →

OL   OL

T-1   T-2   T-3

T-1   T-4

T-3   T-2

120 V connections

T-1   T-4

T-3   T-2

240 V connections

6-45

WIRING DIAGRAM OF HAND OFF MOTOR CONTROL

Note: The controls and the motor operate on the same voltage.

WIRING DIAGRAM OF JOGGING WITH CONTROL RELAY

The jogging circuit allows the motor starter to be energized only when the jog button is depressed.

CONTACTOR AND MOTOR STARTER TROUBLESHOOTING GUIDE

Problem	Possible Cause	Corrective Action
Humming noise	Magnet pole faces misaligned	Realign. Replace magnet assembly if realignment is not possible.
	Too low voltage at coil	Measure voltage at coil. Check voltage rating of coil. Correct any voltage that is 10% less than coil rating.
	Pole face obstructed by foreign object, dirt, or rust	Remove any foreign object and clean as necessary. Never file pole faces.
Loud buzz noise	Shading coil broken	Replace coil assembly.
Controller fails to drop out	Voltage to coil not being removed	Measure voltage at coil. Trace voltage from coil to supply looking for shorted switch or contact if voltage is present.
	Worn or rusted parts causing binding	Clean rusted parts. Replace worn parts.
	Contact poles sticking	Checking for burning or sticky substance on contacts. Replace burned contacts. Clean dirty contacts.
	Mechanical interlock binding	Check to ensure interlocking mechanism is free to move when power is OFF. Replace faulty interlock.
Controller fails to pull in	No coil voltage	Measure voltage at coil terminals. Trace voltage loss from coil to supply voltage if voltage is not present.
	Too low voltage	Measure voltage at coil terminals. Correct voltage level if voltage is less than 10% of rated coil voltage. Check for a voltage drop as large loads are energized.
	Coil open	Measure voltage at coil. Remove coil if voltage is present and correct but coil does not pull in. Measure coil resistance for open circuit. Replace if open.

CONTACTOR AND MOTOR STARTER TROUBLESHOOTING GUIDE (cont.)

Problem	Possible Cause	Corrective Action
Controller fails to pull in	Coil shorted	Shorted coil may show signs of burning. The fuse or breakers should trip if coil is shorted. Disconnect one side of coil and reset if tripped. Remove coil and check resistance for short if protection device does not trip. Replace any coil that is burned.
	Mechanical obstruction	Remove any obstructions.
Contacts badly burned or welded	Too high inrush current	Measure inrush current. Check load for problem if higher-than-load current. Change to larger controller if load current is correct but excessive for controller.
	Too fast load cycling	Change to larger controller if load cycled ON and OFF repeatedly.
	Too large overcurrent protection device	Size overcurrent protection to load and controller.
	Short circuit	Check fuses or breakers. Clear any short circuit.
	Insufficient contact pressure	Check to ensure contacts are making good connection.
Nuisance tripping	Incorrect overload size	Check size of overload against rated load current. Size up if permissible per NEC®.
	Lack of temperature compensation	Correct setting of overload if controller and load are at different ambient temperatures.
	Loose connections	Check for loose terminal connection.

DIRECT CURRENT MOTOR TROUBLESHOOTING GUIDE

Problem	Possible Cause	Corrective Action
Motor will not start	Blown fuse or open CB	Test the OCPD. If voltage is present at the input, but not at the output of the OCPD, the fuse is blown or the CB is open. Check the rating of the OCPD. It should be at least 125% of the motor's FLC.
	Motor overload on starter tripped	Allow overloads to cool. Reset overloads. If reset overloads do not start motor, test the starter.
	No brush contact	Check brushes. Replace, if worn.
	Open control circuit between incoming power and motor	Check for cleanliness, tightness, and breaks. Use a voltmeter to test the circuit, starting with the incoming power and moving to the motor terminals. Voltage generally stops at the problem area.
Fuse, CB, or overloads retrip after service	Excessive load	If the motor is loaded to excess or is jammed, the circuit OCPD will open. Disconnect the load from the motor. If the motor now runs properly, check the load. If the motor does not run and the fuse or CB opens, the problem is with the motor or control circuit. Remove the motor from the control circuit and connect it directly to the power source. If the motor runs properly, the problem is in the control circuit. Check the control circuit. If the motor opens the fuse or CB again, the problem is in the motor. Replace or service the motor.
	Motor shaft does not turn	Disconnect the motor from the load. If the motor shaft still does not turn, the bearings are frozen. Replace or service the motor.
Brushes chip or break	Brush material is too weak or the wrong type for motor's duty rating	Replace with better grade or type of brush. Consult manufacturer if problem continues.
	Brush face is overheating and losing brush bonding material	Check for an overload on the motor. Reduce the load as required. Adjust brush holder arms.

DIRECT CURRENT MOTOR TROUBLESHOOTING GUIDE (cont.)

Problem	Possible Cause	Corrective Action
Brushes chip or break	Brush holder is too far from commutator	Too much space between the brush holder and the surface of the commutator allows the brush end to chip or break. Set correct space between brush holder and commutator.
	Brush tension is incorrect	Adjust brush tension so the brush rides freely on the commutator.
Brushes spark	Worn brushes	Replace worn brushes. Service the motor if rapid brush wear, excessive sparking, chipping, breaking, or chattering is present.
	Commutator is concentric	Grind commutator and undercut mica. Replace commutator if necessary.
	Excessive vibration	Balance armature. Check brushes. They should be riding freely.
Rapid brush wear	Wrong brush material, type, or grade	Replace with brushes recommended by manufacturer.
	Incorrect brush tension	Adjust brush tension so the brush rides freely on the commutator.
Motor overheats	Improper ventilation	Clean all ventilation openings; vacuum or blow dirt out of motor with low-pressure, dry, compressed air.
	Motor is overloaded	Check the load for binding. Check shaft straightness. Measure motor current under operating conditions. If the current is above the listed current rating, remove the motor. Remeasure the current under no-load conditions. If the current is excessive under load but not when unloaded, check the load. If the motor draws excessive current when disconnected, replace or service the motor.

THREE-PHASE MOTOR TROUBLESHOOTING GUIDE

Problem	Possible Cause	Corrective Action
Motor will not start	Wrong motor connections	Most 3φ motors are dual-voltage. Check for proper motor connections.
	Blown fuse or open CB	Test the OCPD. If voltage is present at the input, but not at the output of the OCPD, the fuse is blown or the CB is open. Check the rating of the OCPD. It should be at least 125% of the motor's FLC.
	Motor overload on starter tripped	Allow overloads to cool. Reset overloads. If reset overloads do not start the motor, test the starter.
	Low or no voltage applied to motor	Check the voltage at the motor terminals. The voltage must be present and within 10% of the motor nameplate voltage. If voltage is present at the motor but the motor is not operating, remove the load the motor is driving. Reapply power to the motor. If the motor runs, the problem is with the load. If the motor does not run, the problem is with the motor. Replace or service the motor.
	Open control circuit between incoming power and motor	Check for cleanliness, tightness, and breaks. Use a voltmeter to test the circuit, starting with the incoming power and moving to the motor terminals. Voltage generally stops at the problem area.
Fuse, CB, or overloads retrip after service	Power not applied to all three lines	Measure voltage at each power line. Correct any power supply problems.
	Blown fuse or open CB	Test the OCPD. If voltage is present at the input, but not at the output of the OCPD, the fuse is blown or the CB is open. Check the rating of the OCPD. It should be at least 125% of the motor's FLC.
	Motor overload on starter tripped	Allow overloads to cool. Reset overloads. If reset overloads do not start the motor, test the starter.

THREE-PHASE MOTOR TROUBLESHOOTING GUIDE (cont.)

Problem	Possible Cause	Corrective Action
Fuse, CB, or over-loads retrip after service	Low or no voltage applied to motor	Check the voltage at the motor terminals. The voltage must be present and within 10% of the motor nameplate voltage. If voltage is present at the motor but the motor is not operating, remove the motor from the load the motor is driving. Reapply power to the motor. If the motor runs, the problem is with the load. If the motor does not run, the problem is with the motor. Replace or service the motor.
	Open control circuit between incoming power and motor	Check for cleanliness, tightness, and breaks. Use a voltmeter to test the circuit, starting with the incoming power and moving to the motor terminals. Voltage generally stops at the problem area.
	Motor shaft does not turn	Disconnect the motor from the load. If the motor shaft still does not turn, the bearings are frozen. Replace or service the motor.
Motor overheats	Motor is single phasing	Check each of the 3φ power lines for correct voltage.
	Improper ventilation	Clean all ventilation openings. Vacuum or blow dirt out of motor with low-pressure, dry, compressed air.
	Motor is overloaded	Check the load for binding. Check shaft straightness. Measure motor current under operating conditions. If the current is above the listed current rating, remove the motor. Remeasure the current under no-load conditions. If the current is excessive under load but not when unloaded, check the load. If the motor draws excessive current when disconnected, replace or service the motor.

SPLIT-PHASE MOTOR TROUBLESHOOTING GUIDE

Problem	Possible Cause	Corrective Action
Motor will not start	Thermal cutout switch is open	Reset the thermal switch. Caution: Resetting the thermal switch may automatically start the motor.
	Blown fuse or open CB	Test the OCPD, if voltage is present at the input, but not at the output of the OCPD, the fuse is blown or the CB is open. Check the rating of the OCPD. It should be at least 125% of the motor's FLC.
	Motor overload on starter tripped	Allow overloads to cool. Reset overloads. If reset overloads do not start the motor, test the starter.
	Low or no voltage applied to motor	Check the voltage at the motor terminals. The voltage must be present and within 10% of the motor nameplate voltage. If voltage is present at the motor but the motor is not operating, remove the motor from the load the motor is driving. Reapply power to the motor. If the motor runs, the problem is with the load. If the motor does not run, the problem is with the motor. Replace or service the motor.
	Open control circuit between incoming power and motor	Check for cleanliness, tightness, and breaks. Use a voltmeter to test the circuit, starting with the incoming power and moving to the motor terminals. Voltage generally stops at the problem area.
	Starting winding not receiving power	Check the centrifugal switch to make sure it connects the starting winding when the motor is OFF.
Fuse, CB, or overloads retrip after service	Blown fuse or open CB	Test the OCPD, if voltage is present at the input, but not at the output of the OCPD, the fuse is blown or the CB is open. Check the rating of the OCPD. It should be at least 125% of the motor's FLC.
	Motor overload on starter tripped	Allow overloads to cool. Reset overloads. If reset overloads do not start the motor, test the starter.

SPLIT-PHASE MOTOR TROUBLESHOOTING GUIDE (cont.)

Problem	Possible Cause	Corrective Action
Fuse, CB, or overloads retrip after service	Low or no voltage applied to motor	Check the voltage at the motor terminals. The voltage must be present and within 10% of the motor nameplate voltage. If voltage is present at the motor but the motor is not operating, remove the motor from the load the motor is driving. Reapply power to the motor. If the motor runs, the problem is with the load. If the motor does not run, the problem is with the motor. Replace or service the motor.
	Open control circuit between incoming power and motor	Check for cleanliness, tightness, and breaks. Use a voltmeter to test the circuit, starting with the incoming power and moving to the motor terminals. Voltage generally stops at the problem area.
	Motor shaft does not turn	Disconnect the motor from the load. If the motor shaft still does not turn, the bearings are frozen. Replace or service the motor.
Motor produces electric shock	Broken or disconnected ground strap	Connect or replace ground strap. Test for proper ground.
	Hot power lead at motor connecting terminals is touching motor frame	Disconnect the motor. Open the motor terminal box and check for poor connections, damaged insulation, or leads touching the frame. Service and test motor for ground.
	Motor winding shorted to frame	Remove, service, and test motor.
Motor overheats	Starting windings are not being removed from circuit as motor accelerates	When the motor is turned OFF, a distinct click should be heard as the centrifugal switch closes.
	Improper ventilation	Clean all ventilation openings. Vacuum or blow dirt out of motor with low-pressure, dry, compressed air.

SPLIT-PHASE MOTOR TROUBLESHOOTING GUIDE (cont.)

Problem	Possible Cause	Corrective Action
Motor overheats	Motor is overloaded	Check the load for binding. Check shaft straightness. Measure motor current under operating conditions. If current is above the listed current rating, remove the motor. Remeasure the current under no-load conditions. If the current is excessive under load but not when unloaded, check the load. If the motor draws excessive current when disconnected, replace or service the motor.
	Dry or worn bearings	Dry or worn bearings cause noise. The bearings may be dry due to dirty oil, oil not reaching the shaft, or motor overheating. Oil the bearings as recommended. If noise remains, replace the bearings or the motor.
	Dirty bearings	Clean or replace bearings.
Excessive noise	Excessive end play	Check and play by trying to move the motor shaft in and out. Add end play washers as required.
	Unbalanced motor or load	An unbalanced motor or load causes vibration, which causes noise. Realign the motor and load. Check for excessive end play or loose parts. If the shaft is bent, replace the rotor or motor.
	Dry or worn bearings	Dry or worn bearings cause noise. The bearings may be dry due to dirty oil, oil not reaching the shaft, or motor overheating. Oil the bearings as recommended. If noise remains, replace the bearings or the motor.
	Excessive grease	Ball bearings that have excessive grease may cause the bearings to overheat. Overheated bearings cause noise. Remove any excess grease.

CHAPTER 7
Transformers

SIZING TRANSFORMER CIRCUITS

1. Determine current in primary windings.
   Example: 75kVA, 3-phase, 480/208V, wye-connected transformer. Full-load current = 75,000VA ÷ 480V = 156 amperes ÷ 1.732 = 90.2 amperes per phase.

2. Refer to Table 450.3(B) of the NEC to determine size of overcurrent protective device.
   Example: 90.2 × 125% = 112.75 amperes.

3. Choose next-higher standard overcurrent device.
   Example: 125 amperes, per Section 240.6(A).

4. Size primary conductors per overcurrent device.
   Example: #2 THHN conductors, per Table 310.15(B)(16).

5. Determine if secondary overcurrent protection is required.
   Example: Secondary protection required, per Section 240.21(C)(1).

6. Determine secondary current.
   Example: 75,000VA ÷ 208V = 360.6 amperes ÷ 1.732 = 208 amperes per phase.

7. Refer to Table 450.3(B) of the NEC to determine size of overcurrent protective device.
   Example: 208 × 125% = 260 amperes.

8. Choose next-higher standard overcurrent device.
   Example: 300 amperes, per Section 240.6(A).

9. Size secondary conductors per overcurrent device.
   Example: 300kcmil THHN conductors, per Table 310.15(B)(16).

10. Size main bonding jumper at secondary disconnect, per 250.102(C) and 250.66.
    Example: #2 copper, per Table 250.66.

11. Size grounding electrode conductor.
    Example: #2 copper, per Table 250.66.

12. Choose grounding electrode per Section 250.30(A)(7) and job conditions.

THREE-PHASE CONNECTIONS

Wye Connection

The line current and phase current are the same:

$I_{LINE} = I_{PHASE}$

The line voltage is higher than the phase voltage by a factor of the square root of 3:

$E_{LINE} = E_{PHASE} \times \sqrt{3}$ or $E_{PHASE} = \dfrac{E_{LINE}}{\sqrt{3}}$

Delta Connection

The line voltage and phase voltage are the same:

$E_{LINE} = E_{PHASE}$

The line current is higher than the phase current by a factor of the square root of 3

$I_{LINE} = I_{PHASE} \times \sqrt{3}$ or $I_{PHASE} = \dfrac{I_{LINE}}{\sqrt{3}}$

Open-Delta Connection

Open-delta connections provide 86.6% of the sum of the power rating of two transformers. For example, if you take two transformers rated at 60 kVA each, the total power rating for this connection would be as follows:

60 + 60 = 120 kVA

120 kVA × 0.866 = 103.92 kVA

Voltage and current are computed in the same manner as for a closed-delta connection.

TRANSFORMER TURNS RATIO

A transformer increases or decreases voltage by inducing electrical energy from one coil to another through magnetic lines of force. The turns ratio is the ratio between the voltage and the number of turns on the primary and secondary windings of a transformer.

Thus

$$\frac{N_p}{N_s} = \frac{E_p}{E_s}$$

where

N_s = number of turns in secondary

E_p = primary voltage

E_s = secondary voltage

Transformer with 2:1 Voltage Ratio

480 V   240 V   Primary Voltage = 2 times secondary voltage

100 amps primary   200 amps secondary

20,000 turns   10,000 turns   Secondary Current = 2 times primary current

Primary Winding Turns = 2 Times Secondary Winding Turns

FEEDER TAP AND TRANSFORMER
TAP INSTALLATION GUIDELINES

Feeder conductors shall be permitted to be tapped without overcurrent protection as follows (see NEC 240.21 for all details):

Up to a 10 ft Feeder Tap
- Feeder may be tapped with a smaller conductor.
- Ampacity of tap conductor not less than the combined load, the rating of device supplied, or the rating of the overcurrent protection at end of tap conductors.

Up to a 25 ft Feeder Tap
- Feeder may be tapped with a smaller conductor.
- Ampacity of tap conductor not less than $\frac{1}{3}$ of the rating of the overcurrent protection of feeder.

Conductors shall be permitted to be connected to a transformer secondary without overcurrent protection as follows:

Secondary Conductors up to 10 ft
- Ampacity of conductors not less than the combined load, the rating of device supplied, or the rating of the overcurrent protection at end of secondary conductors.

Secondary Conductors up to 25 ft
- If the conductors on the primary are protected at their ampacity, the conductors on the secondary may be run up to 25 ft before overcurrent protection is needed.

Outdoor Transformer Tap
- There is no limit to the conductor length.
- Single main must be provided where conductors enter a building.

TRANSFORMER FEED TAP NOT OVER 25 FEET LONG

See NEC 240.21 for all details

Feeder panel

25 ft max

Any distance

Any distance

Breaker or set of fuses

25 ft max

225A

225A Bus

208/120 V
208A-FLC.

300MCM TW-(CU)

25 ft max

480 V 90A-FLC.

#2-TW (CU)

7-5

SIZING SINGLE-PHASE TRANSFORMERS

1. Determine the total voltage required by the loads.
2. Determine the kVA capacity required by the loads.
3. Check frequency of supply voltage and the loads.
4. Check supply voltage with rating of primary side.

To calculate kVA capacity of a 1ϕ transformer

$$kVA_{CAP} = E \times \frac{I}{1,000}$$

Where, kVA_{CAP} = transformer capacity (in kVA)

E = voltage (in volts)

I = current (in amps)

The transformer must have a kVA capacity
10% greater than that required by the loads.

kVA	1ϕ FULL-LOAD CURRENTS (AMPS)				
Rating	120 V	208 V	240 V	480 V	2,400 V
1	8.33	4.81	4.17	2.08	.42
3	25.0	14.4	12.5	6.25	1.25
5	41.7	24.0	20.8	10.4	2.08
7.5	62.5	36.1	31.3	15.6	3.13
10	83.3	48.1	41.7	20.8	4.17
15	125.0	72.1	62.5	31.3	6.25
25	208.3	120.2	104.2	52.1	10.4
37.5	312.5	180.3	156.3	78.1	15.6
50	416.7	240.4	208.3	104.2	20.8
75	625.0	360.6	312.5	156.3	31.3
100	833.3	480.8	416.7	208.3	41.7
125	1,041.7	601.0	520.8	260.4	52.1
167.5	1,395.8	805.3	697.9	349.0	69.8
200	1,666.7	961.5	833.3	416.7	83.3
250	2,083.3	1,201.9	1,041.7	520.8	104.2
333	2,775.0	1,601.0	1,387.5	693.8	138.8
500	4,166.7	2,403.8	2,083.3	1,041.7	208.3

SIZING THREE-PHASE TRANSFORMERS

To calculate kVA capacity of a 3ϕ transformer

$$kVA_{CAP} = E \times 1.732 \times \frac{I}{1,000}$$

Where, kVA_{CAP} = transformer capacity (in kVA)

E = voltage (in volts)

1.732 = constant (for 3ϕ power)

I = current (in amps)

kVA	3ϕ FULL-LOAD CURRENTS (AMPS)				
Rating	208 V	240 V	480 V	2,400 V	4,160 V
3	8.3	7.2	3.6	.72	.416
6	16.7	14.4	7.2	1.44	.83
9	25.0	21.7	10.8	2.17	1.25
15	41.6	36.1	18.0	3.6	2.08
30	83.3	72.2	36.1	7.2	4.16
45	124.9	108.3	54.1	10.8	6.25
75	208.2	180.4	90.2	18.0	10.4
100	277.6	240.6	120.3	24.1	13.9
150	416.4	360.9	180.4	36.1	20.8
225	624.6	541.3	270.6	54.1	31.2
300	832.7	721.7	360.9	72.2	41.6
500	1,387.9	1,202.8	601.4	120.3	69.4
750	2,081.9	1,804.3	902.1	180.4	104.1
1,000	2,775.8	2,405.7	1,202.8	240.6	138.8
1,500	4,163.7	3,608.5	1,804.3	360.9	208.2
2,000	5,551.6	4,811.4	2,405.7	481.1	277.6
2,500	6,939.5	6,014.2	3,007.1	601.4	347.0
5,000	13,879.0	12,028.5	6,014.2	1,202.8	694.0
7,500	20,818.5	18,042.7	9,021.4	1,804.3	1,040.9
10,000	27,758.0	24,057.0	12,028.5	2,405.7	1,387.9

TRANSFORMER DERATINGS RESULTING FROM HIGHER AMBIENT TEMPERATURES

Maximum Ambient Temperature (°C) Surrounding Transformer	Maximum Transformer Loading in Percent %
40	100
45	96
50	92
55	88
60	81
65	80
70	76

CALCULATING DERATED kVA CAPACITY FROM HIGHER AMBIENT TEMPERATURES

To calculate the derated kVA capacity of a transformer operating at a higher-than-normal ambient temperature condition, use the formula below:

$$kVA = rated\ kVA \times maximum\ load$$

Where, kVA = derated transformer capacity (in kVA)

rated kVA = manufacturer transformer rating (in kVA)

maximum load = maximum transformer loading (in %)

CURRENT-VOLTAGE RELATIONSHIP BETWEEN HIGH SIDE AND LOW SIDE OF A TRANSFORMER

TESTING A TRANSFORMER FOR POLARITY

DIAGRAM OF A CURRENT TRANSFORMER

SINGLE-PHASE TRANSFORMER CIRCUITS

Series-Connected Winding

Parallel-Connected Winding

**Series-Connected Winding
for 3-Wire Service**

SINGLE-PHASE TRANSFORMER CONNECTIONS

Double-Parallel Connection

Parallel-Series Secondary Connection

7-11

SINGLE-PHASE TRANSFORMER CONNECTIONS *(cont.)*

SINGLE PHASE TO SUPPLY 120 V LIGHTING LOAD

The transformer is connected between high voltage line and load with the 120/240 V winding connected in parallel. This connection is used where the load is comparatively small and the length of the secondary circuit is short. It is often used for a single customer.

SINGLE PHASE TO SUPPLY 120/240 V, 3-WIRE LIGHTING AND POWER LOAD

Here the 120/240 V winding is connected in series and the midpoint brought out, making it possible to serve both 120 and 240 V loads simultaneously. This connection is used in most urban distribution circuits.

SINGLE-PHASE TRANSFORMER CONNECTIONS *(cont.)*

SINGLE PHASE FOR POWER

In this case the 120/240 V winding is connected in series serving 240 V on a two-wire system. This connection is used for small industrial applications.

TWO-PHASE CONNECTIONS

This connection consists merely of two single-phase transformers operated 90° out of phase. For a three-wire secondary as shown, the common wire must carry 32 times the load current. In some cases, a four-wire or a five-wire secondary may be used.

VARIOUS TRANSFORMER CONNECTIONS

SCOTT CONNECTED THREE-PHASE TO TWO-PHASE

When two-phase power is required from a three-phase system, the Scott connection is used the most. The secondary may be three, four, or five wire. Special taps must be provided at 50% and 86.6% of normal primary voltage to make this connection.

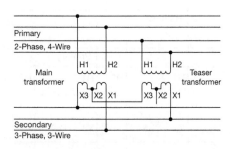

SCOTT CONNECTED TWO-PHASE TO THREE-PHASE

If it should be necessary to supply three-phase power from a two-phase system, the scott connection may be used again. The special taps must be provided on the secondary side, but the connection is similar to the three-phase to two-phase.

To obtain the Scott transformation without a special 86.6% tapped transformer, use one with 10% or two 5% taps to approximate the desired value. A small error of unbalance (overvoltage) occurs that requires care in application.

THREE-PHASE TRANSFORMER CONNECTIONS

OPEN WYE-DELTA

When operating wye-delta with one phase disabled, service may be maintained at reduced load as shown. The neutral in this case must be connected to the neutral of the setup bank through a copper conductor. The system is unbalanced, electro-statically and electro-magnetically, so that telephone interference may be expected if the neutral is connected to ground. The useful capacity of the open delta, open wye bank is 87% of the capacity of the installed transformers when the two units are identical.

DELTA-WYE FOR LIGHTING AND POWER

In the previous banks the single-phase lighting load is all on one phase, resulting in unbalanced primary currents in any one bank. To eliminate this difficulty, the delta-wye system finds many uses. Here the neutral of the secondary three-phase system is grounded and the single-phase loads are connected between the different phase wires and the neutral, while the three-phase loads are connected to the phase wires. Thus, the single-phase load can be balanced on three phases in each bank and banks may be paralleled if desired.

THREE-PHASE TRANSFORMER CONNECTIONS *(cont.)*

WYE-WYE FOR LIGHTING AND POWER

A system in which the primary voltage was increased from 2,400 to 4,160 V to increase the potential capacity. The previously delta-connected distribution transformers are now connected from line to neutral. The secondaries are connected in wye. The primary neutral is connected to the neutral of the supply voltage through a metallic conductor and carried with the phase conductor to minimize telephone interference. If the neutral of the transformer is isolated from the system neutral, an unstable condition results at the transformer neutral caused primarily by third harmonic voltages. If the transformer neutral is connected to ground, the possibility of telephone interference is enhanced with the possibility of resonance between the line capacitance to ground and the magnetizing impedance of the transformer.

WYE-WYE AUTOTRANSFORMERS FOR SUPPLYING POWER FROM A THREE-PHASE, FOUR-WIRE SYSTEM

When ratio of transformation from primary to secondary voltage is small, the best way of stepping down voltage is using autotransformers. It is necessary that the neutral of the autotransformer bank be connected to the system neutral.

THREE-PHASE TRANSFORMER CONNECTIONS *(cont.)*

DELTA-DELTA FOR POWER AND LIGHTING

This connection is used to supply a small single-phase lighting load and three-phase power load simultaneously. As shown, the midtap of the secondary of one transformer is grounded. The small lighting load is connected across the transformer with the midtap and the ground wire common to both 120 V circuits. The single-phase lighting load reduces the available three-phase capacity. This requires special watt-hour metering.

OPEN-DELTA FOR LIGHTING AND POWER

Where the secondary load is a combination of lighting and power, the open-delta connected bank is frequently used. This connection is used when the single-phase lighting load is large as compared with the power load. Here two different size transformers may be used with the lighting load connected across the larger-rated unit.

THREE-PHASE TRANSFORMER CONNECTIONS *(cont.)*

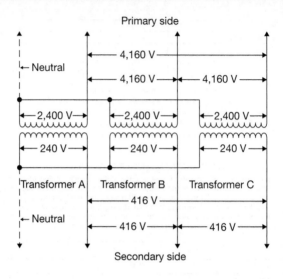

Primary side

4,160 V

4,160 V — 4,160 V

Neutral

2,400 V — 2,400 V — 2,400 V

240 V — 240 V — 240 V

Transformer A | Transformer B | Transformer C

416 V

Neutral

416 V — 416 V

Secondary side

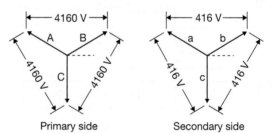

4160 V

A   B

4160 V   4160 V

C

Primary side

416 V

a   b

416 V   416 V

c

Secondary side

Wye-to-Wye Connection

Three 1φ transformers may be connected in a wye-to-wye connection to form a 3φ transformer.

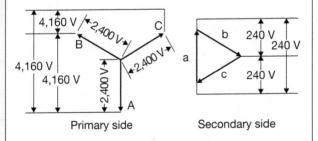

Primary side

Secondary side

Wye-to-Delta Connection

A wye-to-delta connection permits 1ϕ and 3ϕ loads to be drawn simultaneously from the delta-connected secondary at the same voltage.

Primary side

2,400 V

2,400 V — 2,400 V

Transformer A — Transformer B — Transformer C

2,400 V — 2,400 V — 2,400 V

240 V — 240 V — 240 V

416 V — 416 V

416 V

Neutral

1φ transformers connected in
delta on primary side and wye
on secondary side

Secondary side

2,400 V

2,400 V

2,400 V

Primary side

240 V

240 V

240 V

416 V

416 V

416 V

Secondary side

Delta-to-Wye Connection

A 3φ delta-to-wye connection is often used for distribution where a four-wire secondary distribution circuit is required.

THREE-PHASE TRANSFORMER CONNECTIONS *(cont.)*

Line current is √3 x coil
current in a delta connection

Current and voltage values on a delta bank of three
single-phase transformer windings. A voltage of 240
volts and a current of 10 amperes are assumed.

Two Transformers in an Open Delta Arrangement

Wye Connection
Line Volts = 1.732 × Coil Volts
Coil Amps = Line Amps

Current and voltage values on a wye bank of three single-phase transformer windings. A voltage of 120 volts and a current of 10 amperes are assumed.

A SPLIT-COIL TRANSFORMER

AN ADDITIVE-POLARITY TRANSFORMER

7-23

SECONDARY TIES

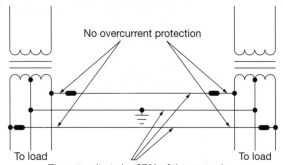

No overcurrent protection

To load

To load

Tie ampacity to be 67% of the secondary
current of the largest transformer.

LOADS CONNECTED BETWEEN
TRANSFORMER SUPPLY POINTS

No overcurrent protection

To load

To load

Tie ampacity to be 100% of the secondary
current of the largest transformer.

AUTOTRANSFORMER

TYPICAL BOOST TRANSFORMER CONNECTIONS

115 V INPUT
H_1 H_2
H_3 H_4
10% BOOST
X_2 X_4 X_1 X_3
OUTPUT

115 V INPUT
H_1 H_2
H_3 H_4
20% BOOST
X_4 X_3 X_2 X_1
OUTPUT

213 V – 227 V INPUT
H_1 H_2 H_3 H_4
5% BOOST
X_2 X_4 X_1 X_3
OUTPUT

205 V – 212 V INPUT
H_1 H_2 H_3 H_4
10% BOOST
X_4 X_3 X_2 X_1
OUTPUT

230 V INPUT
H_1 H_2 H_3 H_4
5% BOOST
X_2 X_4 X_1 X_3
OUTPUT

230 V INPUT
H_1 H_2 H_3 H_4
10% BOOST
X_4 X_3 X_2 X_1
OUTPUT

Note: Always verify with the manufacturer's instructions

7-26

TYPICAL BUCK TRANSFORMER CONNECTIONS

115 V
INPUT

H_1 H_2
H_3 H_4

10% BUCK

X_2 X_1
X_4 X_3

OUTPUT

---

115 V
INPUT

H_1 H_2
H_3 H_4

20% BUCK

X_4 X_3 X_2 X_1

OUTPUT

---

230 V
INPUT

H_1 H_2 H_3 H_4

5% BUCK

X_2 X_1
X_4 X_3

OUTPUT

---

230 V
INPUT

H_1 H_2 H_3 H_4

10% BUCK

X_4 X_3 X_2 X_1

OUTPUT

---

240 V –
250 V   INPUT

H_1 H_2 H_3 H_4

5% BUCK

X_2 X_1
X_4 X_3

OUTPUT

---

251 V –
265 V   INPUT

H_1 H_2 H_3 H_4

10% BUCK

X_4 X_3 X_2 X_1

OUTPUT

Note: Always verify with the manufacturer's instructions

7-27

TRANSFORMER SOUND LEVELS

kVA Rating	Sound Level (in dB)	Hearing Intensity	Example of Loudness
0–5	40	Barely Audible	Refrigerator running
6–9	40		
10–25	45		
26–50	45		
51–150	50	Quiet	Whisper
151–225	55		
226–300	55		
301–500	60	Moderate	Normal conversation

TRANSFORMING WINDING RATIOS

Tap Winding	Primary Voltage	Secondary Voltage	Ratio
Full Winding	2,400	120/240	20:1/10:1
$4^{1}/_{2}$%	2,292	120/240	19:1/9.5:1
9%	2,184	120/240	18:1/9:1
$13^{1}/_{2}$%	2,076	120/240	17:1/8.5:1

SATISFACTORY VOLTAGE LEVELS

Devices	Tolerable Under Voltage	Satisfactory Voltage	Tolerable Over Voltage
Lights, Motors, etc.	112	117	122–127
Stoves, Dryers, etc.	230	235	245–250

CHAPTER 8
Communications and Electronics

TELEPHONE CONNECTIONS			
Typical Inside Wire			
Type of Wire	**Pair No.**	**Pair Color Matches**	
2-pair Wire	1 2	Green Black	Red Yellow
3-pair Wire	1 2 3	White/Blue White/Orange White/Green	Blue/White Orange/White Green/White

Inside Wire Connecting Terminations			
Wire Color		**Wire Function**	
2-pair Wire	3-pair Wire	Service w/o Dial Light	Service with Dial Light
Green Red Black Yellow	White/Blue Blue/White White/Orange Orange/White	Tip Ring Not Used Ground	Tip Ring Transformer Transformer

Typical Fasteners and Recommended Spacing Distances			
Fasteners	**Horizontal**	**Vertical**	**From Corner**
Wire clamp	16.0 in	16.0 in	2 in
Staples (wire)	7.5 in	7.5 in	2 in
Bridle Rings*	4.0 ft	–	2–8.5 in*
Drive Rings**	4.0 ft	8.0 ft	2–8.5 in*

*When changing direction, space fasteners to hold wire at 45-degree angle.
**To avoid injury do not use drive rings below a 6 foot clearance, use bridle rings.

STANDARD TELECOM COLOR CODING

PAIR #	TIP (+) COLOR	RING (−) COLOR
1	White	Blue
2	White	Orange
3	White	Green
4	White	Brown
5	White	Slate
6	Red	Blue
7	Red	Orange
8	Red	Green
9	Red	Brown
10	Red	Slate
11	Black	Blue
12	Black	Orange
13	Black	Green
14	Black	Brown
15	Black	Slate
16	Yellow	Blue
17	Yellow	Orange
18	Yellow	Green
19	Yellow	Brown
20	Yellow	Slate
21	Violet	Blue
22	Violet	Orange
23	Violet	Green
24	Violet	Brown
25	Violet	Slate

SEPARATION AND PHYSICAL PROTECTION FOR PREMISES WIRING

Table applies only to telephone wiring from the network interface modular jacks to telephone equipment. Separations apply to crossing and to parallel runs (minimum separations).

Types of Wire Involved		Minimum Separations	Alternatives
Electric Supply	Bare light or power wire of any voltage	5 ft	No Alternatives
	Open wiring not over 300 V	2 in	*
Radio & TV	Antenna lead-in and grds.	4 in	*
Signal or Control Wires	Open wiring or wires in conduit or cable	None	N/A
Comm. Wires	Coaxial cables w/shield	None	N/A
Telephone Drop Wire	Using fused protectors	2 in	*
	Using fuseless protectors		None
Sign	Neon signs and wiring		6 in
Lightning Sys.	Lightning rods and wires		6 ft

*If minimum separations cannot be obtained, additional protection required.

MINIMUM SEPARATION DISTANCE FROM POWER SOURCE AT 480 V OR LESS

CONDITION	<2 kVA	2–5 kVA	>5 kVA
Unshielded power lines or equipment near open or nonmetal pathways	5 in	12 in	24 in
Unshielded power lines or equipment near a grounded metal conduit pathway	2.5 in	6 in	12 in
Power lines enclosed in a grounded metal conduit (or equivalent)	—	6 in	12 in
Transformers and electric motors	40 in	40 in	40 in
Fluorescent lighting	12 in	12 in	12 in

CONDUCTORS ENTERING BUILDINGS

If on the same pole, or run parallel in span:

1. Communications conductors must be located below power conductors.
2. Communications conductors cannot be connected to crossarms.
3. Power drops must be separated from comm. drops by 12 inches.

Above roofs, communications conductors must have the following clearances:

1. Flat roofs: 8 feet.
2. Garages and other auxiliary buildings: None required.
3. Overhangs, where no more than 4 feet of communications cable will run over the area: 18 inches.
4. Where the roof slope is 4 inches rise for every 12 inches horizontally: 3 feet.

Underground communications conductors must be separated from power conductors in manhole or handholes by brick, concrete, or tile partitions.

Communications conductors should be kept at least 6 feet away from lightning protection system conductors.

CIRCUIT PROTECTION REQUIREMENTS

1. **Protectors** are surge arresters and are required for all aerial circuits not confined with a *block*. They must be installed on all circuits with a block that could accidentally contact power circuits over 300 volts to ground.
2. **Metal sheaths** of any communications cables must be grounded or interrupted with an insulating joint as close as practicable to the point where they enter any building (where the cable emerges through an exterior wall or concrete floor slab, or from a grounded rigid or intermediate metal conduit).
3. **Grounding conductors** for communications circuits must be copper or some other corrosion-resistant material, and have insulation suitable for the area in which it is installed.
4. **Grounding conductors** may be no smaller than 14 AWG.
5. The **grounding conductor** must be run as directly as possible to the grounding electrode, and be protected if necessary.
6. If the **grounding conductor** is protected by metal raceway, it must be bonded to the grounding conductor on both ends.

CIRCUIT PROTECTION REQUIREMENTS *(cont.)*

Grounding electrodes for communications ground may be any of the following:

1. The grounding electrode of an electrical power system
2. A grounded interior metal piping system
3. Metal power service raceway
4. Power service equipment enclosures
5. A separate grounding electrode
6. Any acceptable power system grounding electrode
7. A grounded metal structure
8. A ground rod or pipe at least 5 feet long and 1/2 inch in diameter (This rod should be driven into damp (if possible) earth and kept separate.)

If the power and communications systems use separate grounding electrodes, bind them together with 6 AWG copper. Other electrodes may also be bound.

For mobile homes, if there is no service equipment or disconnect within 30 feet of the mobile home wall, the communications circuit must have its own grounding electrode. In this case, or if the mobile home is connected with cord and plug, the communications circuit protector must be bonded to the mobile home frame or grounding terminal with a copper conductor no smaller than 12 AWG.

INTERIOR COMMUNICATIONS CONDUCTORS

Must be kept at least 2 inches away from power or Class 1 conductors, unless permanently separated from them or they are enclosed in one of the following:

1. Raceway
2. Type AC, MC, UF, or NM cable, or metal-sheathed cable

Communications cables are allowed in the same raceway, box, or cable with:

1. Class 2 and 3 remote-control, signaling, and power-limited circuits.
2. Power-limited fire-protective signaling systems.
3. Conductive or nonconductive optical fiber cables.
4. Community antenna television and radio distribution systems.

Communications conductors are not allowed to be in the same raceway or fitting with power or Class 1 circuits.

Communications conductors are not allowed to be supported by raceways unless the raceway runs directly to the piece of equipment the communications circuit serves.

TELEPHONE CONNECTIONS

Telephone cable contains four wires, colored green, red, black, and yellow. A one-line telephone requires only two wires to operate. Green and red are the conductors used. In a four-wire modular connector, the green and red conductors are in the inside positions, the black and yellow wires in the outer positions. For two-line phones, the inside wires (green and red) carry line 1, and the outside wires (black and yellow) carry line 2.

EIA COLOR CODE

Pair 1 – White/Blue (white with blue stripe) and Blue
Pair 2 – White/Orange and Orange
Pair 3 – White/Green and Green
Pair 4 – White/Brown and Brown

TWISTED-PAIR PLUGS AND JACKS

There are three major types of twisted-pair jacks:
RJ-type connectors (phone plugs)
Pin-connectors
Genderless connectors (IBM sexless data connectors)
The RJ-type (registered jack) refers to the standard used for most telephone jacks. Pin-connector refers to twisted pairs such as the RS-232 connector, which have male and female pin receptacles. Genderless connectors have no separate male or female component.

PHONE JACKS

A **four-conductor jack**, supporting two twisted-pairs, is the one used for connecting telephone handsets to receivers.

A **six-conductor jack**, supporting three twisted-pairs, is the RJ-11 jack used to connect telephones to the telephone company. An **eight-conductor jack** is the R-45 jack, which is intended for use as the user-site interface for ISDN terminals.

These types of jacks are often keyed, so that the wrong type of plug cannot be inserted into the jack.

CROSS CONNECTIONS

Cross connections are made at **terminal blocks**. A block is typically a rectangular, white plastic unit with metal connection points. The most common type is called a punchdown block. Connections are made by pushing the insulated wires into their places. When "punched" down, the connector cuts through the insulation and makes the appropriate connection.

Connections are made between punch-down blocks by using **patch cords**, which are short lengths of cable that can be terminated into the punch-down slots, or that are equipped with connectors on each end.

When different systems must be connected together, cross-connects are used.

CATEGORY CABLING

Category 1 cable is the old standard type of telephone cable, with four conductors.
Category 2 Obsolete.
Category 3 cable is used for digital voice and data transmission rates up to 10 Mbit/s (Megabits per second). Common types of data transmission over this communications cable would be UTP Token Ring (4 Mbit/s) and 10Base-T (10 Mbit/s).
Category 4 Obsolete.
Category 5 cable is used for sending voice and data at speeds up to 100 Mbit/s.
Category 6 cable is used for sending data at speeds up to 200 or 250 Mbit/s.

INSTALLATION REQUIREMENTS

Article 800 of the NEC® covers communication circuits, such as telephone systems and outside wiring for fire and burglar alarm systems. Generally these circuits must be separated from power circuits and grounded. In addition, all such circuits that run out of doors (even if only partially) must be provided with circuit protectors (surge or voltage suppressors).

TYPES OF DATA NETWORKS

10Base2	Ethernet running over thin, 50 ohm baseband coaxial cable at 10 Mbit/s. 10Base2 is also commonly referred to as thin-Ethernet.
10BaseT	Ethernet over twisted-pair at 10 Mbit/s
100BaseTX	Ethernet over twisted-pair at 100 Mbit/s
1000BaseT	Ethernet over twisted-pair at 1 Gbit/s
10GBaseT	Ethernet over twisted-pair at 10 Gbit/s
10GBaseR	Ethernet over 50 micron fiber at 10 Gbit/s

STANDARD CONFIGURATIONS FOR
SEVERAL COMMON NETWORKS

ATM 155 Mbps	pairs 2 and 4 (pins 1-2, 7-8)
Ethernet 10BaseT	pairs 2 and 3 (pins 1-2, 3-6)
Ethernet 100BaseT4	pairs 2 and 3 (4T+) (pins 1-2, 3-6)
Ethernet 100BaseT8	pairs 1, 2, 3, and 4 (pins 4-5, 1-2, 3-6, 7-8)
Token-Ring	pairs 1 and 3 (pins 4-5, 3-6)
TP-PMD	pairs 2 and 4 (pins 1-2, 7-8)
100VG-AnyLAN	pairs 1, 2, 3, and 4 (pins 4-5, 1-2, 3-6, 7-8)

ETHERNET 10BaseT, STRAIGHT-THRU PATCH CORD

RJ45 Plug		RJ45 Plug
T2 White/Orange	1	TxData +
R2 Orange	2	TxData −
T3 White/Green	3	RecvData +
R1 Blue	4	
T1 White/Blue	5	
R3 Green	6	RecvData −
T4 White/Brown	7	
R4 Brown	8	

ETHERNET 10BaseT, CROSSOVER PATCH CORD

This cable is used to cascade hubs, or for connecting two
Ethernet stations back-to-back without a hub.

RJ45 Plug	RJ45 Plug
1 Tx+	Rx+ 3
2 Tx−	Rx− 6
3 Rx+	Tx+ 1
6 Rx−	Tx− 2

EIA-606 COLORS AND FUNCTIONS OF CABLING

Blue	Horizontal voice cables
Brown	Inter-building backbone
Gray	Second-level backbone
Green	Network connections and auxiliary circuits
Orange	Demarcation point
Purple	First-level backbone
Red	Key-type telephone systems
Silver or White	Horizontal data cables, computer, and PBX
Yellow	Auxiliary, maintenance, and security alarms

THERMOSTAT WIRING GUIDE

- Red wire to the white wire turns furnace on.
- Red wire to the green wire turns fan on.
- Red wire to the yellow wire turns A/C on.

or

- **White Wire = Heat On**
- **Green Wire = Fan On**
- **Yellow Wire = A/C On**

WIRING GUIDE FOR THERMOSTAT, FURNACE, AND AIR CONDITIONING

For thermostats with a 2-stage setup, for 2 furnaces and 2 compressors, remove the jumper between RC and RH. For 1-stage, leave the jumper in place.

BASIC THERMOSTAT CIRCUIT

Heating element

Magnetic switch

Thermostat

M

M

Low-voltage circuit

Transformer

115 or 230 V supply lines

TWIN-TYPE THERMOSTAT CIRCUIT

Twin thermostat

Night   Day

W   Y   Y   B   R

Clock

Transformer

Line, 120 V

Primary control

R
B
W

INTRUDER ALARM CIRCUIT WITH ANNUNCIATOR AND RELAY

Neutral

Hot

Master switch

110 V contact

Relay

Alarm bell

Battery

Relay (500 watts or less)

Reset button

Annunciator drops

Drop contacts

Test switches

Door/window springs

8-12

ELECTRONIC WIRING COLOR CODES

Electronic Applications

Color	Circuit Type
Black	Chassis grounds, returns, primary leads
Blue	Plate leads, transistor collectors, FET drain
Brown	Filaments, plate start lead
Gray	AC main power leads
Green	Transistor base, finish grid, diodes, FET gate
Orange	Transistor base 2, screen grid
Red	B plus DC power supply
Violet	Power supply minus
White	B – C minus of bias supply, AVC – AGC return
Yellow	Emitters-cathode and transistor, FET source

Stereo Audio Channels

Color	Circuit Type
White	Left channel high side
Blue	Left channel low side
Red	Right channel high side
Green	Right channel low side

AF Transformers (Audio)

Color	Circuit Type
Black	Ground line
Blue	Plate, collector, or drain lead; end of primary winding
Brown	Start primary loop; opposite to blue lead
Green	High side, end secondary loop
Red	B plus, center tap push-pull loop
Yellow	Secondary center tap

IF Transformers (Intermediate Frequency)

Color	Circuit Type
Blue	Primary high side of plate, collector, or drain lead
Green	Secondary high side for output
Red	Low side of primary returning B plus
Violet	Secondary outputs
White	Secondary low side

COLOR CODES FOR RESISTORS

Radial Lead Resistor

Axial Lead Resistor

Color	1st # (D1)	2nd # (D2)	Multiplier (M)	Tolerance (T)
No Color	–	–	–	20%
Silver	–	–	0.01	10%
Gold	–	–	0.1	5%
White	9	9	10^9	–
Gray	8	8	100,000,000	–
Violet	7	7	10,000,000	–
Blue	6	6	1,000,000	–
Green	5	5	100,000	–
Yellow	4	4	10,000	–
Orange	3	3	1,000	4%
Red	2	2	100	3%
Brown	1	1	10	2%
Black	0	0	1	1%

Example: Yellow – Blue – Brown – Silver = 460 ohms

If only Band D1 is wide, it indicates that the resistor is wirewound. If Band D1 is wide and there is also a blue fifth band to the right of Band T on the Axial Lead Resistor, it indicates the resistor is wirewound and flame proof.

STANDARD VALUES FOR RESISTORS

Values for 5% class   k = kilohms = 1,000 ohms   m = megohms = 1,000,000 ohms

1	8.2	68	560	4.7 k	39 k	330 k	2.7 m
1.1	9.1	75	620	5.1 k	43 k	360 k	3.0 m
1.2	10	82	680	5.6 k	47 k	390 k	3.3 m
1.3	11	91	750	6.2 k	51 k	430 k	3.6 m
1.5	12	100	820	6.8 k	56 k	470 k	3.9 m
1.6	13	110	910	7.5 k	62 k	510 k	4.3 m
1.8	15	120	1.0 k	8.2 k	68 k	560 k	4.7 m
2.0	16	130	1.1 k	9.1 k	75 k	620 k	5.1 m
2.2	18	150	1.2 k	10 k	82 k	680 k	5.6 m
2.4	20	160	1.3 k	11 k	91 k	750 k	6.2 m
2.7	22	180	1.5 k	12 k	100 k	820 k	6.8 m
3.0	24	200	1.6 k	13 k	110 k	910 k	7.5 m
3.3	28	220	1.8 k	15 k	120 k	1.0 m	8.2 m
3.6	30	240	2.0 k	16 k	130 k	1.1 m	9.1 m
3.9	33	270	2.2 k	18 k	150 k	1.2 m	10.0 m
4.3	36	300	2.4 k	20 k	160 k	1.3 m	
4.7	39	330	2.7 k	22 k	180 k	1.5 m	
5.1	43	360	3.0 k	24 k	200 k	1.6 m	
5.6	47	390	3.3 k	27 k	220 k	1.8 m	
6.2	51	430	3.6 k	30 k	240 k	2.0 m	
6.8	56	470	3.9 k	33 k	270 k	2.2 m	
7.5	62	510	4.3 k	36 k	300 k	2.4 m	

CAPACITORS

Mica Capacitor

Ceramic Capacitor

Disc Capacitor

Ceramic disc capacitors are normally labeled. If the number is less than 1 then the value is in picofarads, if greater than 1 the value is in microfarads. The letter R can be used in place of a decimal (e.g., 2R9=2.9).

COLOR CODES FOR CAPACITORS

Color	1st # (D1)	2nd # (D2)	Multiplier (M)	Tolerance (T)
No Color	–	–	–	20%
Silver	–	–	0.01	10%
Gold	–	–	0.1	5%
White	9	9	10^9	9%
Gray	8	8	100,000,000	8%
Violet	7	7	10,000,000	7%
Blue	6	6	1,000,000	6%
Green	5	5	100,000	5%
Yellow	4	4	10,000	4%
Orange	3	3	1,000	3%
Red	2	2	100	2%
Brown	1	1	10	1%
Black	0	0	1	20%

COLOR CODES FOR CERAMIC CAPACITORS

Color	Tolerance (T) Above 10pf	Tolerance (T) Below 10pf	Decimal Multiplier (M)	Temp Coef ppm/°C (C)
White	10	1.0	0.1	500
Gray	–	0.25	0.01	30
Violet	–	–	–	−750
Blue	–	–	–	−470
Green	5	0.5	–	−330
Yellow	–	–	–	−220
Orange	–	–	1000	−150
Red	2	–	100	−80
Brown	1	–	10	−30
Black	20	2.0	1	0

CHARACTERISTICS OF LEAD–ACID BATTERIES

TEMPERATURE (F) VERSUS BATTERY EFFICIENCY (%)

−20° = 18%	20° = 58%
−10° = 33%	30° = 64%
0° = 40%	50° = 82%
10° = 50%	80° = 100%

CHARGE	SPECIFIC GRAVITY OF ACID
Discharged	1.11 to 1.12
Very Low Capacity	1.13 to 1.15
25% of Capacity	1.15 to 1.17
50% of Capacity	1.20 to 1.22
75% of Capacity	1.24 to 1.26
100% of Capacity	1.26 to 1.28
Overcharged	1.30 to 1.32

MAGNETIC PERMEABILITY OF SOME COMMON MATERIALS

Substance	Permeability (approx.)
Aluminum	Slightly more than 1
Bismuth	Slightly less than 1
Cobalt	60–70
Ferrite	100–300
Free space	1
Iron	60–100
Iron, refined	3,000–8,000
Nickel	50–60
Permalloy	3,000–30,000
Silver	Slightly less than 1
Steel	300–600
Super permalloys	100,000–1,000,000
Wax	Slightly less than 1
Wood, dry	Slightly less than 1

TRANSISTOR CIRCUIT ABBREVIATIONS

Quantity	Abbreviations
Base-emitter voltage	E_B, V_B, E_{BE}, V_{BE}
Collector-emitter voltage	E_C, V_C, E_{CE}, V_{CE}
Collector-base voltage	E_{BC}, V_{BC}, E_{CB}, V_{CB}
Gate-source voltage	E_G, V_G, E_{GS}, V_{GS}
Drain-source voltage	E_D, V_D, E_{DS}, V_{DS}
Drain-gate voltage	E_{DG}, V_{DG}, E_{DG}, V_{DG}
Emitter current	I_E
Base current	I_B, I_{BE}, I_{EB}
Collector current	I_C, I_{CE}, I_{EC}
Source current	I_S
Gate current	I_G, I_{GS}, I_{SG}*
Drain current	I_D, I_{DS}, I_{SD}
*This is almost always significant.	

RADIO FREQUENCY CLASSIFICATIONS

Classification	Abbreviation	Frequency range
Very Low Frequency	VLF	9 kHz and below
Low Frequency (Longwave)	LF	30 kHz–300 kHz
Medium Frequency	MF	300 kHz–2MHz
High Frequency (Shortwave)	HF	3 MHz–30MHz
Very High Frequency	VHF	30 MHz–300 MHz
Ultra High Frequency	UHF	300 MHz–3 GHz
Microwaves	MW	3 GHz and more

SMALL TUBE FUSES

CHARACTERISTICS	Fuse Type	Fuse Diameter	Fuse Length
Dual Element, time delay, glass tube	MDL	0.250"	1.250"
Dual Element, glass tube	MDX	0.250"	1.250"
Dual Element, glass tube, pigtail	MDV	0.250"	1.250"
Ceramic Body, normal, 200% 15 sec	3AB	0.250"	1.250"
Metric, fast blow, high int., 210% 30 min	216	5 mm	20 mm
Glass, metric, fast blow, 210% 30 min	217	5 mm	20 mm
Glass, metric, slow blow, 210% 2 min	218	5 mm	20 mm
No Delay, ceramic, 110% rating, opens at 135% load in 1 hr	ABC	0.250"	1.250"
Fast Acting, glass tube, 110% rating, opens at 135% load in 1 hr	AGC	0.250"	1.250"
Fast Acting, glass tube	AGX	0.250"	1.000"
No Delay, 200% 15 sec	BLF	0.406"	1.500"
No Delay, military, 200% 15 sec	BLN	0.406"	1.500"
Fast Cleaning, 600V, 135% 1 hr	BLS	0.406"	1.375"
Time Delay, indicator pin, 135% 1 hr	FLA	0.406"	1.500"
Dual Element, delay, 200% 12 sec	FLM	0.406"	1.500"
Dual Element, delay, 500V, 200% 12 sec	FLQ	0.406"	1.500"
Slow Blow, time delay	FNM	0.406"	1.250"
Slow Blow, indicator, metal pin pops outs indicating blown, dual element	FNA	0.406"	1.500"
Rectifier Fuse, fast, low let through	GBB	0.250"	1.250"
Indicator Fuse, metal pin pops out indicating blown, 110% rating	GLD	0.250"	1.250"
Metric, fast acting	GGS	5 mm	20 mm
Fast, current limiting, 600V, 135% 1 hr	KLK	0.406"	1.500"
Fast, protect solid state, 250% 1 sec	KLW	0.406"	1.500"
Slow Blow, time delay size rejection also	SC	0.406"	1.625" to 2.250"
Slow Blow, glass body, 200% 5 sec	218000	0.197"	0.787"
Slow Blow, glass body, 200% 5 sec	313000	0.250"	1.250"
Slow Blow, ceramic, 200% 5 sec	326000	0.250"	1.250"
Auto Glass, fast blow, 200% 5 sec	1AG	0.250"	0.625"
Auto Glass, fast blow, 200% 10 sec	2AG	0.177"	0.570"
Auto Glass, fast blow, 200% 5 sec	3AG	0.250"	1.250"
Auto Glass, fast blow, 200% 5 sec	4AG	0.281"	1.250"
Auto Glass, fast blow, 200% 5 sec	8AG	0.406"	1.500"
Auto Glass, fast blow, 200% 5 sec	7AG	0.250"	0.875"
Auto Glass, fast blow, 200% 5 sec	8AG	0.250"	1.000"
Auto Glass, fast blow, 200% 5 sec	9AG	0.250"	1.438"

The percentage and time figures mean, for example, that a 135% overload will blow a KLK type fuse in 1 hour.

CHAPTER 9
Electrical and Job Safety

ELECTRICAL SAFETY GUIDELINES

- Always comply with the NEC®.
- Use UL-approved appliances, components, and equipment.
- Keep electrical grounding circuits in good condition. Ground any conductive component or element that does not have to be energized. The grounding connection must be a low-resistance conductor heavy enough to carry the largest fault current that may occur.
- Turn OFF, lock out, and tag disconnect switches when working on any electrical circuit or equipment. Test all circuits after they are turned OFF.
- Use double-insulated power tools or power tools that include a third conductor grounding terminal, which provides a path for fault current.
- Always use protective and safety equipment.
- Check conductors, cords, components, and equipment for signs of wear or damage.
- Never throw water on an electrical fire. Turn OFF the power and use a Class C–rated fire extinguisher.
- Never work alone when working in a dangerous area or with dangerous equipment.
- Learn CPR and first aid.
- Do not work in poorly lighted areas.

ELECTRICAL SAFETY GUIDELINES *(cont.)*

- Always use nonconductive ladders, never metal.
- Ensure there are no atmospheric hazards such as flammable dust or vapor in the work area.
- Use one hand when working on a live circuit to reduce the chance of an electrical shock passing through the heart and lungs.
- Never bypass or disable fuses or circuit breakers.
- Put in place an effective lockout/tagout system.
- Promptly replace frayed, damaged, or worn electrical cords or cables.

EFFECT OF ELECTRIC CURRENT ON THE HUMAN BODY

Current (ma)	Effect	Result
0 to 6	Slight sensation possible	None
6 to 15	Painful shock muscular contraction	Possible No "Let Go"
15 to 20	Painful shock Frozen until circuit is de-energized	No "Let Go"
20 to 50	Severe muscular contractions Asphyxia	Often Fatal
50 to 200	Ventricular fibrillation	Probably Fatal
200 +	Heart movement stops	Fatal

ELECTRICAL SAFETY CHECKLIST

- GFCI-protected receptacles, cords, etc.
- Prohibit work on new and existing energized (hot) electrical circuits until all power is shut off and grounds are attached.
- Ensure that all extension cords have grounding prongs.
- Protect flexible cords and cables from damage.
- Use extension cord sets that are three-wire type and imprinted: S, ST, SO, STO.
- Maintain all electrical tools and equipment in safe condition and check them regularly for defects.
- Do not bypass any protective system or device.
- Locate and identify overhead power lines.
- Ensure that ladders never come within 10 feet (3 m) of electrical power lines.
- Assume that all overhead wires are energized at lethal voltages. Never assume a wire is safe to touch if it is down or appears insulated.
- Never touch a fallen overhead power line.
- Stay at least 10 feet (3 m) away from overhead wires during cleanup and other activities.
- If an overhead wire falls across your vehicle while driving, stay inside and continue to drive away from the line. If the engine stalls, do not leave vehicle. Call or ask someone to call emergency services.
- Never operate electrical equipment while you are standing in water.
- If working in damp locations, inspect electric cords and equipment and use a ground-fault circuit interrupter.
- Always use caution when working near electricity.

LOCKOUT/TAGOUT PROCEDURES

SHUTDOWN AND LOCKOUT/TAGOUT

1. Notify all affected personnel that work to be performed requires lockout/tagout. Only authorized personnel are permitted to lockout equipment.
2. Prepare for shutdown by reviewing company procedures and identifying all energy sources. Locks and tags shall be uniform and easily identified and only used to lockout equipment for personal protection.
3. Shut down equipment by normal means and isolate the equipment from all energy sources. No one other than the person who installed the lockout tag-out device shall attempt to operate any switch, valve, or isolating device bearing a lock or lockout tag.
4. Lockout and tag all energy disconnects or isolating devices. Use individual locks. Primary sources of energy include electricity and pressurized gases or fluids.
5. Release all sources of static or stored energy including mechanical motion, springs, capacitors, thermal energy, and residual pressure in gas or fluid systems.
6. Verify the isolation by operating the normal operating controls.

LOCK REMOVAL AND RETURN TO SERVICE

1. Check that all tools have been removed, all equipment components are intact, and that all employees are completely clear of the equipment.
2. Verify that all controls are in a neutral position so as to prevent any type of mechanical engagement.
3. Remove all lockout devices and re-energize the equipment. A device may only be removed by the authorized employee who applied it.
4. Notify all affected personnel that the work has been completed and the equipment is back in operation.

WORK AREA CONTROL

- A barrier tape, cones, fence, or other methods may be used to surround any work area where a hazard exists that could be accessed by personnel unfamiliar with that work, such as onlookers.
- The tape shall be prominently placed and completely surround the area where the hazard exists.
- The taped area shall be large enough to permit all persons working within the work area adequate clearance from any hazard.
- Remove all barriers when the hazard is eliminated.
- Metal barrier tape shall not be used.
- There may be a need for attendants to provide additional safety for personnel protection. This may include the use of personnel with flags, detour signs, cones, etc.

OSHA SAFETY COLOR CODES

Color	Examples
Red	Fire protection equipment and apparatus; portable containers of flammable liquids; emergency stop pushbuttons/switches
Yellow	Caution and for marking physical hazards; waste containers for explosive or combustible materials; caution against starting, using, or moving equipment under repair; identifies the starting point or power source of machinery
Orange	Dangerous parts of machines; safety starter buttons; the exposed parts of pulleys, gears, rollers, cutting, and power jaws
Purple	Radiation hazards
Green	Safety areas and location of first aid equipment

TOP OSHA ELECTRICAL VIOLATIONS (2001)

	Section	Description	Total Violations
1	.404(b)(1)	Electrical—Ground-fault protection	758
2	.405(a)(2)	Electrical—Wiring methods—Temporary wiring	495
3	.404(f)(6)	Electrical—Wiring design and protection—Grounding path	346
4	.405(g)(2)	Electrical—Identification, splices, and terminations	304
5	.403(b)(2)	Electrical—Examination, installation, and use of equipment	223
6	.403(i)(2)	Electrical—Guarding electrical equipment live parts	174
7	.416(e)(1)	Electrical—Safety-related work practices—Cords and cables	162
8	.405(b)(2)	Electrical—Wiring methods, components, and equipment	139
9	.405(b)(1)	Electrical—Wiring methods—Cabinets, boxes, and fittings	128
10	.403(b)(1)	Electrical—Approved equipment	117
11	.416(a)(1)	Electrical—Protection of employees	80
12	.405(j)(1)	Electrical—Lighting fixtures, lamps, and receptacles	78
13	.405(g)(1)	Electrical—Use of flexible cords and cables	69

WIRING DESIGN AND PROTECTION

Branch Circuits – Ground-Fault Protection

(i) General. The employer shall use either ground-fault circuit interrupters as specified in paragraph (b)(1)(ii) of this section or an assured equipment grounding conductor program as specified in paragraph (b)(1)(iii) of this section to protect employees on construction sites. These requirements are in addition to any other requirements for equipment grounding conductors.

(ii) Ground-fault circuit interrupters. All 120-volt, single-phase, 15- and 20-ampere receptacle outlets on construction sites, which are not a part of the permanent wiring of the building or structure and are in use by employees, shall have approved ground-fault circuit interrupters for personnel protection. Receptacles on a two-wire, single-phase portable or vehicle-mounted generator rated not more than 5 kW, where the circuit conductors of the generator are insulated from the generator frame and all other grounded surfaces, need not be protected with ground-fault circuit interrupters.

(iii) Assured equipment-grounding conductor program. The employer shall establish and implement an assured equipment-grounding conductor program on construction sites covering all cord sets, receptacles that are not a part of the building or structure and equipment connected by cord and plug that are available for use or used by employees. This program shall comply with the following minimum requirements:

(A) A written description of the program, including the specific procedures adopted by the employer, shall be available at the job site for inspection and copying by the assistant secretary and any affected employee.

(B) The employer shall designate one or more competent persons to implement the program.

(C) Each cord set, attachment cap, plug and receptacle of cord sets, and any equipment connected by cord and plug, except cord sets and receptacles that are fixed and not exposed to damage, shall be visually inspected before each day's use for external defects, such as deformed or missing pins or insulation damage, and for indications of possible internal damage. Equipment found to be damaged or defective shall not be used until repaired.

(D) The following tests shall be performed on all cord sets, receptacles that are not a part of the permanent wiring of the building or structure and cord- and plug-connected equipment required to be grounded:

(1) All equipment-grounding conductors shall be tested for continuity and shall be electrically continuous.

(2) Each receptacle and attachment cap or plug shall be tested for correct attachment of the equipment grounding conductor. The equipment-grounding conductor shall be connected to its proper terminal.

(E) All required tests shall be performed:

(1) Before first use;

(2) Before equipment is returned to service following any repairs;

(3) Before equipment is used after any incident that can be reasonably suspected to have caused damage (e.g., when a cord set is run over); and

(4) At intervals not exceeding 3 months, but cord sets and receptacles that are fixed and not exposed to damage shall be tested at intervals not exceeding 6 months.

(F) The employer shall not make available or permit the use by employees of any equipment that has not met the requirements of paragraph (iii) of this section.

(G) Tests performed as required in this paragraph shall be recorded. This test record shall identify each receptacle, cord set, and cord-and plug-connected equipment that passed the test and shall indicate the last date on which it was tested or the interval for which it was tested. This record shall be kept by means of logs, color coding, or other effective means and shall be maintained until replaced by a more current record. The record shall be made available on the job site for inspection by the assistant secretary and any affected employee.

REQUIRED WORKING CLEARANCE — MINIMUM DEPTH IN FEET

Nominal Voltage to Ground	Working Conditions		
	1	2	3
0 to 150 V	3'	3'	3'
151 to 600 V	3'	4'	4'
601 to 2,500 V	3'	4'	5'
2,501 to 9,000 V	4'	5'	6'
9,001 to 25,000 V	5'	6'	9'
25,001 to 75,000 V	6'	8'	10'
Over 75,000 V	8'	10'	12'

Working Conditions:

Condition 1 – Exposed live parts on one side and no live or grounded parts on the other side. Exposed live parts on both sides effectively guarded by suitable insulating materials.

Condition 2 – Exposed live parts on one side only. Grounded parts on the other side. Concrete, brick, or tile walls are considered grounded.

Condition 3 – Exposed live parts on both sides (not guarded).

MINIMUM ELEVATION OF UNGUARDED LIVE PARTS ABOVE WORKING SPACE IN FEET

Nominal Voltage between Phases	Elevation
601 to 7,500 V	9'
7,501 to 35,000 V	9½'
Over 35,000 V	9½' + 0.37" for every 1,000 volts over 35,000 volts

MINIMUM CLEARANCE — LIVE PARTS

| *Minimum Clearance of Live Parts (in) | | | | Nominal Voltage Rating (kV) | Impulse Withstand B.I.L. (kV) | |
| Phase-to-Phase | | Phase-to-Ground | | | | |
Indoors	Outdoors	Indoors	Outdoors		Indoors	Outdoors
4.5	7	3.0	6	2.4–4.16	60	95
5.5	7	4.0	6	7.2	75	95
7.5	12	5.0	7	13.8	95	110
9.0	12	6.5	7	14.4	110	110
10.5	15	7.5	10	23	125	150
12.5	15	9.5	10	34.5	150	150
18.0	18	13.0	13	—	200	200
—	18	—	13	46	—	200
—	21	—	17	—	—	250
—	21	—	17	69	—	250
—	31	—	25	—	—	350
—	53	—	42	115	—	550
—	53	—	42	138	—	550
—	63	—	50	—	—	650
—	63	—	50	161	—	650
—	72	—	58	—	—	750
—	72	—	58	230	—	750
—	89	—	71	—	—	900
—	105	—	83	—	—	1,050

*This represents the minimum clearance for rigid parts and bare conductors under favorable conditions. They shall be increased for conductor movement, unfavorable conditions, or where space is limited. Impulse withstand voltage for a particular system voltage is determined by the type of surge protection equipment that is utilized.

PORTABLE LADDER SAFETY

- Read and follow all labels and markings.

- Avoid electrical hazards! Look for overhead power lines before handling a ladder. Never use a metal ladder near power lines.

- Always inspect the ladder before using it.

- Always maintain a three point (two hands and a foot, or two feet and a hand) contact on the ladder when climbing.

- Only use ladders and appropriate accessories for their designated purposes.

- Ladders must be free of any slippery material on the rungs, steps, or feet.

- Do not use a self-supporting ladder as a single ladder or in a partially closed position.

- Do not use the top step/rung of a ladder.

- Use a ladder only on a stable and level surface.

- Never place a ladder on a box, barrel, or other unstable base to obtain additional height.

- Never move or shift a ladder while a person or equipment is on it.

- An extension or straight ladder used to access an elevated surface must extend at least 3 feet above the point of support.

- Never stand on the top three rungs of a straight, single, or extension ladder.

- The proper angle for setting up a ladder is to place its base a quarter of the working length of the ladder from the wall or other vertical surface.

PORTABLE LADDER SAFETY *(cont.)*

- A ladder placed in any location where it can be displaced by other work activities must be secured, or a barricade must be erected around it.
- Be sure that all locks are properly engaged.
- Never exceed the maximum load rating of a ladder. Be aware of the weight it is supporting (including tools).

ELECTRICAL GLOVE INSPECTION

- Gloves shall be inspected before and after every use.
- Check that the voltage rating of the insulating equipment is greater than the application voltage.
- Check for defects by stretching a small area at a time.
- Look for cracks, cuts, scratches, embedded material, or weak spots.
- Inspect the entire glove. Any detectable flaw is cause for gloves to be returned for testing.
- Check for pinhole leaks by trapping air in the fingers of the glove. Also squeeze the cuff with one hand and inspect the palm, thumb, and fingers for defects.
- Hold the glove to your face to check for air leakage.
- Any leakage is cause for the glove to be destroyed. Cut the glove in half immediately.
- Check the test date stamped (month and year) on any rubber protective equipment.
- Maximum time between tests shall be 6 months for gloves and 12 months for sleeves and insulating blankets.

TREE TRIMMING AND REMOVAL SAFETY

Assume All Power Lines Are Energized!

- Contact the utility company to discuss de-energizing and grounding or shielding of power lines.
- All tree trimming and removal work within 10 feet of a power line must be done by trained and experienced line-clearance tree trimmers. A second tree trimmer is required within normal voice communication range.
- Line-clearance tree trimmers must be aware of and maintain the proper minimum approach distances when working around energized power lines.
- Use extreme caution when moving ladders and equipment around downed trees and power lines.

Stay Alert at All Times!

- Do not trim trees in dangerous weather conditions.
- Perform a hazard assessment of the work area before starting work.
- Eliminate or minimize exposure to hazards at the tree and in the surrounding area.
- Operators of chainsaws and other equipment should be trained and the equipment properly maintained.
- Use personal protective equipment such as gloves, safety glasses, hard hats, hearing protection, etc., as recommended in the equipment manufacturers' operating manuals.

PORTABLE GENERATOR SAFETY

Major Causes of Injuries and Fatalities
- Shocks and electrocution from improper use or accidentally energizing other equipment.
- Carbon monoxide from a generator's exhaust.
- Fires from improperly refueling the generator.

Safe Work Practices
- Inspect portable generators for damage or loose fuel lines.
- Keep the generator dry.
- Maintain and operate portable generators in accordance with the manufacturer's manual.
- Never attach a generator directly to the electrical system of a structure unless the generator has a properly installed transfer switch.
- Always plug electrical appliances directly into the generator using the manufacturer-supplied cords. Use undamaged heavy-duty extension cords that are grounded (three-pronged).
- Use ground-fault circuit interrupters.
- Before refueling, shut down the generator. Never store fuel indoors.

Carbon Monoxide Poisoning
Many people have died from CO poisoning because their generator was not adequately ventilated.
- Never use a generator indoors.
- Never place a generator outdoors near doors, windows, or vents.
- If you or others show these symptoms—dizziness, headaches, nausea, tiredness—get to fresh air immediately and seek medical attention.

FIRST AID GUIDE

FIRST — CHECK THE SCENE AND THE INJURED PERSON.
THEN — CALL 911 OR THE LOCAL EMERGENCY NUMBER.
THEN — CARE FOR THE INJURED PERSON.

BLEEDING

1) Direct Pressure and Elevation —
   - Place dressing and apply pressure directly over wound, then elevate above the level of the heart unless there is evidence of a fracture.

2) Apply Pressure Bandage —
   - Wrap bandage tightly over the dressing.

3) Pressure Points —
   - If bleeding doesn't stop after direct pressure, elevation, and the pressure bandage, compress the pressure point.
   - Arm — Push the brachial artery against the upper arm bone.
   - Leg — Apply pressure on femoral artery, pushing it against the femur or pelvic bone.

4) Nosebleed —
   - Be sure to have the victim lean forward and pinch the nostrils together until bleeding stops.

POISONING

Signs: Vomiting, heavy labored breathing, sudden onset of pain or illness, burns or odor around the lips or mouth, extreme or bizarre behavior.

- **Call the poison control center or local emergency number and follow their directions.**
- Try to identify the poison and inform the poison control center of the type, when incident occurred, victim's age, size and symptoms, and quantity of poison that may have been ingested, inhaled, absorbed, or injected.

If unconscious or nauseous

- Position victim on side and monitor vital signs.
- **DO NOT** give anything by mouth.

FIRST AID GUIDE *(cont.)*

SHOCK
Signs: Cool, moist, pale, bluish skin, weak rapid pulse (100+), increased breathing rate, lethargy, and nausea.

SHOCK CARE
1) Have the victim lie down in a comfortable position to minimize pain, which can intensify the body's stress and accelerate the progression of shock.
2) Control any external bleeding.
3) Maintain normal body temperature.
4) Elevate legs a foot unless there are possible head, neck, back injuries, or broken bones involving the hips or legs. If unsure, leave the victim lying flat.
5) Do not give anything by mouth.
6) Call 911 or your local emergency number at once.

BURNS
Signs: *For thin burns:* Redness, pain, and swelling.
For deep burns: Blisters, deep tissue destruction, and charred appearance.
1) Put out the heat — Remove the victim from the source of the burn.
2) Cool all burns (except electrical) — Run or pour cool water on burn. Immerse if possible.
3) Cover the burn — Use dry, sterile dressing and bandage.
4) Keep victim comfortable — Not chilled or overheated.

Chemical burns: Must be flushed with large amounts of water until help arrives.

Electrical burn: Be absolutely sure power is **OFF** before touching the victim.

ELECTRICAL SHOCK
Signs: Unconsciousness, absence of breathing, and/or pulse.
1) **TURN OFF THE POWER SOURCE AND CALL 911.**
2) **DO NOT** approach the victim until power de-energized.
3) **DO NOT** move victim unless necessary.
4) Administer **CPR** if necessary.
5) Treat for shock and check for other injuries.

FIRST AID GUIDE *(cont.)*

FROSTBITE

Signs: Flushed, white or gray skin. Pain may be felt early, then subside. Blisters may appear.

1) Cover the frozen part. Loosen clothing or footwear.
2) Give the victim a warm drink. (No alcohol or caffeine.)
3) Immerse frozen part in warm water up to 105°, or wrap in warm blankets.
4) Remove from water and discontinue once part becomes flushed.
5) Elevate the area.
6) **DO NOT** rub the frozen part. **DO NOT** break the blisters. **DO NOT** use extreme or dry heat to rewarm the part.
7) If fingers or toes are involved, place gauze between them when bandaging.

HYPOTHERMIA

Signs: Lowered body core temperature. Persistent shivering, lips blue, slow slurred speech, and memory lapses.

1) Move victim to shelter and remove any wet clothing.
2) Rewarm victim with blankets or body-to-body contact.
3) Give warm liquids if possible and keep victim quiet.
4) **DO NOT** give alcohol or caffeine.
5) Monitor victim and give **CPR** if necessary.

HEAT EXHAUSTION/HEAT STROKE

Signs: *Heat Exhaustion:* Pale, clammy skin, profuse perspiration, weakness, nausea, and headache.
Heat Stroke. Hot dry red skin, no perspiration, rapid pulse. High body temperature (105°+).

1) Call 911.
2) Get the victim out of the heat.
3) Loosen tight or restrictive clothing.
4) Remove perspiration-soaked clothing.
5) Apply cool, wet cloths to the skin and fan victim.
6) Give cool water if possible.

FIRST AID GUIDE *(cont.)*

NOT BREATHING

1. CHECK THE VICTIM.

If no response . . .

2. CALL 911.

3. CARE FOR THE VICTIM.

Look, listen, and feel for breathing for 5 to 7 seconds.

If the person is not breathing

Step 1: Position victim on back and support head and neck.

Step 2: Tilt head back and lift chin.

Step 3: Look, listen, and feel for breathing for 5 seconds.

STILL NOT BREATHING

Step 1: Give two slow gentle breaths.

Step 2: Check pulse for 5 to 7 seconds.

Step 3: Check for profuse bleeding.

4. BEGIN RESCUE BREATHING.

If pulse is present but person is still not breathing . . .

Step 1: Give one slow breath every five seconds for one minute.

Step 2: Recheck pulse and breathing every minute.

Step 3: **Continue rescue breathing as long as pulse is present but person is not breathing.**

5. BEGIN CPR IF NO SIGN OF PULSE AND NO BREATHING.

CHOKING

1. Check the victim.

Step 1: Ask, **"Are you choking?"** If victim cannot respond, breathe, is coughing weakly, or making noises . . .

Step 2: **Phone 911.**

Step 3: **Do Abdominal Thrusts.**

    A. Wrap your arms around victim's waist. Make a fist. Place thumbside of fist against middle of abdomen just above the navel. Grasp fist with other hand.

    B. Give quick, upward thrusts.

Repeat until object is dislodged or person becomes unconscious.

TYPES OF FIRE EXTINGUISHERS

TYPE A: To extinguish fires involving trash, cloth, paper, and other wood- or pulp-based materials. The flames are put out by water-based ingredients or dry chemicals.

TYPE B: To extinguish fires involving greases, paints, solvents, gas, and other petroleum-based liquids. The flames are put out by cutting off oxygen and stopping the release of flammable vapors. Dry chemicals, foams, and halon are used.

TYPE C: To extinguish fires involving electricity. The combustion is put out the same way as with a type B extinguisher but, most importantly, the chemical in a type C and <u>MUST</u> be nonconductive to electricity in order to be safe and effective.

TYPE D: To extinguish fires involving combustible metals. Please be advised to obtain important information from your local fire department on the requirements for type D fire extinguishers for your area.

Any combination of letters indicates that an extinguisher will put out more than one type of fire. Type BC will put out two types of fires. The size of the fire to be extinguished is shown by a number in front of the letter, such as 100A. For example,

Class 1A will extinguish 25 burning sticks 40 inches long.

Class 1B will extinguish a paint thinner fire 2.5 square feet in size.

Class 100B will put out a fire 100 times larger than type 1B.

Here are some basic guidelines to follow:

- By using a type ABC you will cover most basic fires.

- Use fire extinguishers with a gauge and that are constructed with metal. Also note if the unit is UL approved.

- Utilize more than one extinguisher and be sure that each unit is mounted in a clearly visible and accessible manner.

- After purchasing any fire extinguisher, always review the basic instructions for its intended use. Never deviate from the manufacturer's guidelines. Following this simple procedure could end up saving lives.

CHAPTER 10
Materials and Tools

STRENGTH GAIN VS. PULL ANGLE

The weight bearing capacity of a strap, for example, increases by the factor K (shown below) as the angle of the strap decreases. A 100 lb capacity strap at a 60° pulling angle loses 50 lb of weight-bearing ability. At perfectly vertical, the ability is 100%.

K	D°	K	D°	K	D°
.7412	75	.3572	50	.0937	25
.6580	70	.2929	45	.0603	20
.5774	65	.2340	40	.0341	15
.5000	60	.1340	35	.0152	10
.4264	55	.1208	30	.0038	5

LENGTH OF WIRE CABLE PER REEL

Cable length (feet) = X (X + Y) Z (K) Use equation above to determine the length of wire cable that is wound smoothly on a reel. The dimensions for X, Y, and Z are in inches.

(K)	Cable Dia. (in)	(K)	Cable Dia. (in)
.0476	2.250	.239	1.000
.0532	2.125	.308	.875
.0597	2.000	.428	.750
.0675	1.875	.607	.625
.0770	1.750	.741	.563
.0886	1.625	.925	.500
.107	1.500	1.19	.438
.127	1.375	1.58	.375
.152	1.250	2.21	.313
.191	1.125	3.29	.250

SHEET METAL SCREW CHARACTERISTICS

Screw Size #	Screw Dia. (in)	Diameter of Pierced Hole (in)	Hole Size #	Thickness of Metal – Gauge #
4	.112	.086	44	28
		.086	44	26
		.093	42	24
		.098	42	22
		.100	40	20
6	.138	.111	39	28
		.111	39	26
		.111	39	24
		.111	38	22
		.111	36	20
7	.155	.121	37	28
		.121	37	26
		.121	35	24
		.121	33	22
		.121	32	20
		—	31	18
8	.165	.137	33	26
		.137	33	24
		.137	32	22
		.137	31	20
		—	30	18
10	.191	.158	30	26
		.158	30	24
		.158	30	22
		.158	29	20
		.158	25	18
12	.218	—	26	24
		.185	25	22
		.185	24	20
		.185	22	18
14	.251	—	15	24
		.212	12	22
		.212	11	20
		.212	9	18

Deviations in materials and conditions could require variations from these dimensions.

WOOD SCREW CHARACTERISTICS

Screw Size #	Wood Screw Standard Lengths (in)	Size of Pilot Hole		Size of Shank Hole	
		Softwood Bit #	Hardwood Bit #	Clearance Bit #	Hole Diameter (in)
0	1/4	75	66	52	.060
1	1/4 to 3/8	71	57	47	.073
2	1/4 to 1/2	65	54	42	.086
3	1/4 to 5/8	58	53	37	.099
4	3/8 to 3/4	55	51	32	.112
5	3/8 to 3/4	53	47	30	.125
6	3/8 to 1 1/2	52	44	27	.138
7	3/8 to 1 1/2	51	39	22	.151
8	1/2 to 2	48	35	18	.164
9	5/8 to 2 1/4	45	33	14	.177
10	5/8 to 2 1/2	43	31	10	.190
11	3/4 to 3	40	29	4	.203
12	7/8 to 3 1/2	38	25	2	.216
14	1 to 4 1/2	32	14	D	.242
16	1 1/4 to 5 1/2	29	10	I	.268
18	1 1/2 to 6	26	6	N	.294
20	1 3/4 to 6	19	3	P	.320
24	3 1/2 to 6	15	D	V	.372

HEX HEAD BOLT AND TORQUE CHARACTERISTICS

BOLT MAKE-UP IS STEEL WITH COARSE THREADS

Number of Threads per Inch	Hex Head Bolt Size (in)	Torque = Foot-Pounds		
		⬡ SAE 0-1-2 74,000 psi	⬡ SAE Grade 3 100,000 psi	⬡ SAE Grade 5 120,000 psi
4.5	2	2,750	5,427	4,550
5	1¾	1,900	3,436	3,150
6	1½	1,100	1,943	1,775
6	1⅜	900	1,624	1,500
7	1¼	675	1,211	1,105
7	1⅛	480	872	794
8	1	310	551	587
9	⅞	206	372	382
10	¾	155	234	257
11	⅝	96	145	154
12	9⁄16	69	103	114
13	½	47	69	78
14	7⁄16	32	47	54
16	⅜	20	30	33
18	5⁄16	12	17	19
20	¼	6	9	10

For fine thread bolts, increase by 9%.

HEX HEAD BOLT AND TORQUE CHARACTERISTICS (cont.)

BOLT MAKE-UP IS STEEL WITH COARSE THREADS

Number of Threads per Inch	Hex Head Bolt Size (in)	Torque = Foot-Pounds		
		SAE Grade 6 133,000 psi	SAE Grade 7 133,000 psi	SAE Grade 8 150,000 psi
4.5	2	7,491	7,500	8,200
5	1¾	5,189	5,300	5,650
6	1½	2,913	3,000	3,200
6	1⅜	2,434	2,500	2,650
7	1¼	1,815	1,825	1,975
7	1⅛	1,304	1,325	1,430
8	1	825	840	700
9	⅞	550	570	600
10	¾	350	360	380
11	⅝	209	215	230
12	⁹⁄₁₆	150	154	169
13	½	106	110	119
14	⁷⁄₁₆	69	71	78
16	⅜	43	44	47
18	⁵⁄₁₆	24	25	29
20	¼	12.5	13	14

For fine thread bolts, increase by 9%.
For special alloy bolts, obtain torque rating from the manufacturer.

BOLT MAKE-UP IS STEEL WITH COARSE THREADS

Number of Threads per Inch	Whitworth Type Hex Head Bolt Size (in)	Grades A & B 62,720 psi	Grade S 112,000 psi	Grade T 123,200 psi	Grade V 145,600 psi
			Torque = Foot-Pounds		
8	1	276	497	611	693
9	7/8	186	322	407	459
11	3/4	118	213	259	287
11	5/8	73	128	155	175
12	9/16	52	94	111	128
12	1/2	36	64	79	89
14	7/16	24	43	51	58
16	3/8	15	27	31	36
18	5/16	9	15	18	21
20	1/4	5	7	9	10

For fine thread bolts, increase by 9%.

METRIC HEX HEAD BOLT AND TORQUE CHARACTERISTICS

BOLT MAKE-UP IS STEEL WITH COARSE THREADS

Torque = Foot-Pounds

Thread Pitch (mm)	Bolt Size (mm)	Standard 5D 71,160 psi	Standard 8G 113,800 psi	Standard 10K 142,000 psi	Standard 12K 170,674 psi
3.0	24	261	419	570	689
2.5	22	182	284	394	464
2.0	18	111	182	236	183
2.0	16	83	132	175	208
1.25	14	55	89	117	137
1.25	12	34	54	70	86
1.25	10	19	31	40	49
1.0	8	10	16	22	27
1.0	6	5	6	8	10

For fine thread bolts, increase by 9%.

ALLEN HEAD AND MACHINE SCREW BOLT AND TORQUE CHARACTERISTICS

Number of Threads per Inch	Allen Head and Mach. Screw Bolt Size	Torque in Foot-Pounds or Inch-Pounds		
		Allen Head Case H Steel 160,000 psi	Mach. Screw Yellow Brass 60,000 psi	Mach. Screw Silicone Bronze 70,000 psi
4.5	2"	8,800	–	–
5	1¾"	6,100	–	–
6	1½"	3,450	655	595
6	1⅜"	2,850	–	–
7	1¼"	2,130	450	400
7	1⅛"	1,520	365	325
8	1"	970	250	215
9	⅞"	640	180	160
10	¾"	400	117	104
11	⅝"	250	88	78
12	⁹⁄₁₆"	180	53	49
13	½"	125	41	37
14	⁷⁄₁₆"	84	30	27
16	⅜"	54	20	17
18	⁵⁄₁₆"	33	125 in#	110 in#
20	¼"	16	70 in#	65 in#
24	#10	60	22 in#	20 in#
32	#8	46	19 in#	16 in#
32	#6	21	10 in#	8 in#
40	#5	–	7.2 in#	6.4 in#
40	#4	–	4.9 in#	4.4 in#
48	#3	–	3.7 in#	3.3 in#
56	#2	–	2.3 in#	2 in#

For fine thread bolts, increase by 9%.

TIGHTENING TORQUE IN POUND-FEET-SCREW FIT			
Wire Size, AWG/kcmil	Driver	Bolt	Other
18–16	1.67	6.25	4.2
14–8	1.67	6.25	6.125
6–4	3.0	12.5	8.0
3–1	3.2	21.00	10.40
0–2/0	4.22	29	12.5
3/0–200	–	37.5	17.0
250–300	–	50.0	21.0
400	–	62.5	21.0
500	–	62.5	25.0
600–750	–	75.0	25.0
800–1,000	–	83.25	33.0
1,250–2,000	–	83.26	42.0

SCREW TORQUES

Screw Size in Inches Across, Hex Flats	Torque in Pound-Feet
1/8	4.2
5/32	8.3
3/16	15
7/32	23.25
1/4	42

STANDARD TAPS AND DIES

Thread Size	Coarse			Fine		
	Drill Size	Threads per Inch	Decimal Size	Drill Size	Threads per Inch	Decimal Size
4"	3	4	3.75"	–	–	–
3¾"	3	4	3.5"	–	–	–
3½"	3	4	3.25"	–	–	–
3¼"	3	4	3.0"	–	–	–
3"	2	4	2.75"	–	–	–
2¾"	2	4	2.5"	–	–	–
2½"	2	4	2.25"	–	–	–
2¼"	2	4.5	2.0313"	–	–	–
2"	1	4.5	1.7813"	–	–	–
1¾"	1	2	1.5469"	–	–	–
1½"	1	6	1.3281"	1²⁷⁄₆₄"	12	1.4219"
1⅜"	1	6	1.2188"	1¹⁹⁄₆₄"	12	1.2969"
1¼"	1	7	1.1094"	1¹¹⁄₆₄"	12	1.1719"
1⅛"	63⁄64"	7	.9844"	1³⁄₆₄"	12	1.0469"
1"	⅞"	8	.8750"	¹⁵⁄₁₆"	14	.9375"
⅞"	49⁄64"	9	.7656"	¹³⁄₁₆"	14	.8125"
¾"	21⁄32"	10	.6563"	¹¹⁄₁₆"	16	.6875"
⅝"	17⁄32"	11	.5313"	³⁷⁄₆₄"	18	.5781"
9⁄16"	31⁄64"	12	.4844"	³³⁄₆₄"	18	.5156"
½"	27⁄64"	13	.4219"	²⁹⁄₆₄"	20	.4531"
7⁄16"	U	14	.368"	²⁵⁄₆₄"	20	.3906"
⅜"	5⁄16"	16	.3125"	Q	24	.332"
5⁄16"	F	18	.2570"	I	24	.272"
¼"	#7	20	.201"	#3	28	.213"
#12	#16	24	.177"	#14	28	.182"
#10	#25	24	.1495"	#21	32	.159"
3⁄16"	#26	24	.147"	#22	32	.157"
#8	#29	32	.136"	#29	36	.136"
#6	#36	32	.1065"	#33	40	.113"
#5	#38	40	.1015"	#37	44	.104"
⅛"	3⁄32"	32	.0938"	#38	40	.1015"
#4	#43	40	.089"	#42	48	.0935"
#3	#47	48	.0785"	#45	56	.082"
#2	#50	56	.07"	#50	64	.07"
#1	#53	64	.0595"	#53	72	.0595"
#0	–	–	–	3⁄64"	80	.0469"

Thread Pitch	TAPS AND DIES — METRIC CONVERSIONS			
	Fine Thread Size		Tap Drill Size	
(mm)	in	mm	in	mm
4.5	1.6535	42	1.4567	37.0
4.0	1.5748	40	1.4173	36.0
4.0	1.5354	39	1.3779	35.0
4.0	1.4961	38	1.3386	34.0
4.0	1.4173	36	1.2598	32.0
3.5	1.3386	34	1.2008	30.5
3.5	1.2992	33	1.1614	29.5
3.5	1.2598	32	1.1220	28.5
3.5	1.1811	30	1.0433	26.5
3.0	1.1024	28	.9842	25.0
3.0	1.0630	27	.9449	24.0
3.0	1.0236	26	.9055	23.0
3.0	.9449	24	.8268	21.0
2.5	.8771	22	.7677	19.5
2.5	.7974	20	.6890	17.5
2.5	.7087	18	.6102	15.5
2.0	.6299	16	.5118	14.0
2.0	.5512	14	.4724	12.0
1.75	.4624	12	.4134	10.5
1.50	.4624	12	.4134	10.5
1.50	.3937	11	.3780	9.6
1.50	.3937	10	.3386	8.6
1.25	.3543	9	.3071	7.8
1.25	.3150	8	.2677	6.8
1.0	.2856	7	.2362	6.0
1.0	.2362	6	.1968	5.0
.90	.2165	5.5	.1811	4.6
.80	.1968	5	.1653	4.2
.75	.1772	4.5	.1476	3.75
.70	.1575	4	.1299	3.3
.75	.1575	4	.1279	3.25
.60	.1378	3.5	.1142	2.9
.60	.1181	3	.0945	2.4
.50	.1181	3	.0984	2.5
.45	.1124	2.6	.0827	2.1
.45	.0984	2.5	.0787	2.0
.40	.0895	2.3	.0748	1.9
.40	.0787	2	.0630	1.6
.45	.0787	2	.0590	1.5
.35	.0590	1.5	.0433	1.1

RECOMMENDED DRILLING SPEEDS

Material	Bit Sizes	RPM Speed Range	
Glass	Special Metal Tube Drilling	700 Only	
Plastics	7/16" and larger	500	– 1,000
	3/8"	1,500	– 2,000
	5/16"	2,000	– 2,500
	1/4"	3,000	– 3,500
	3/16"	3,500	– 4,000
	1/8"	5,000	– 6,000
	1/16" and smaller	6,000	– 6,500
Woods	1" and larger	700	– 2,000
	3/4" to 1"	2,000	– 2,300
	1/2" to 3/4"	2,300	– 3,100
	1/4" to 1/2"	3,100	– 3,800
	1/4" and smaller	3,800	– 4,000
	carving / routing	4,000	– 6,000
Soft Metals	7/16" and larger	1,500	– 2,500
	3/8"	3,000	– 3,500
	5/16"	3,500	– 4,000
	1/4"	4,500	– 5,000
	3/16"	5,000	– 6,000
	1/8"	6,000	– 6,500
	1/16" and smaller	6,000	– 6,500
Steel	7/16" and larger	500	– 1,000
	3/8"	1,000	– 1,500
	5/16"	1,000	– 1,500
	1/4"	1,500	– 2,000
	3/16"	2,000	– 2,500
	1/8"	3,000	– 4,000
	1/16" and smaller	5,000	– 6,500
Cast Iron	7/16" and larger	1,000	– 1,500
	3/8"	1,500	– 2,000
	5/16"	1,500	– 2,000
	1/4"	2,000	– 2,500
	3/16"	2,500	– 3,000
	1/8"	3,500	– 4,500
	1/16" and smaller	6,000	– 6,500

TORQUE LUBRICATION EFFECTS IN FOOT-POUNDS

Lubricant	5/16" - 18 Thread	1/2" - 13 Thread	Torque Decrease
Graphite	13	62	49–55%
Mily Film	14	66	45–52%
White Grease	16	79	35–45%
Sae 30	16	79	35–45%
Sae 40	17	83	31–41%
Sae 20	18	87	28–38%
Plated	19	90	26–34%
No Lube	29	121	0%

METALWORKING LUBRICANTS

Materials	Threading	Lathing	Drilling
Machine Steels	Dissolvable Oil Mineral Oil Lard Oil	Dissolvable Oil	Dissolvable Oil Sulpherized Oil Min. Lard Oil
Tool Steels	Lard Oil Sulpherized Oil	Dissolvable Oil	Dissolvable Oil Sulpherized Oil
Cast Irons	Sulpherized Oil Dry Min. Lard Oil	Dissolvable Oil Dry	Dissolvable Oil Dry Air Jet
Malleable Irons	Soda Water Lard Oil	Soda Water Dissolvable Oil	Soda Water Dry
Aluminums	Kerosene Dissolvable Oil Lard Oil	Dissolvable Oil	Kerosene Dissolvable Oil
Brasses	Dissolvable Oil Lard Oil	Dissolvable Oil	Kerosene Dissolvable Oil Dry
Bronzes	Dissolvable Oil Lard Oil	Dissolvable Oil	Dissolvable Oil Dry
Coppers	Dissolvable Oil Lard Oil	Dissolvable Oil	Kerosene Dissolvable Oil Dry

TYPES OF SOLDERING FLUX

To solder	Use
Cast iron Galvanized iron, galvanized, steel Brass, copper, gold, iron, silver, steel	Cuprous oxide Hydrochloric acid Borax
Bronze, cadmium, copper, lead Brass, copper, gun metal, nickel, tin, zinc Bismuth, brass, copper, gold, silver, tin	Resin Ammonia chloride Zinc chloride
Silver Pewter and Lead Stainless	Sterling Tallow Stainless steel (only)

HARD SOLDER ALLOYS

To Hard Solder	Copper %	Gold %	Silver %	Zinc %
Gold	22	67	11	–
Silver	20	–	70	10
Hard brass	45	–	–	55
Soft brass	22	–	–	78
Copper	50	–	–	50
Cast iron	55	–	–	45
Steel and iron	64	–	–	36

SOFT SOLDER ALLOYS

To Soft Solder	Lead %	Tin %	Zinc %	Bism %	Other %
Gold	33	67	–	–	–
Silver	33	67	–	–	–
Brass	34	66	–	–	–
Copper	40	60	–	–	–
Steel and iron	50	50	–	–	–
Galvanized steel	42	58	–	–	–
Tinned steel	36	64	–	–	–
Zinc	45	55	–	–	–
Block Tin	1	99	–	–	–
Lead	67	33	–	–	–
Gun metal	37	63	–	–	–
Pewter	25	25	–	50	–
Bismuth	33	33	–	34	–
Aluminum	–	70	25	–	5

PROPERTIES OF WELDING GASES

Type of Gas	Characteristics	Common Tank Sizes (cu ft)
Acetylene	C_2H_2, explosive gas, flammable, garlic-like odor, colorless, dangerous if used in pressures over 15 psig (30 psig absolute)	10, 40, 75 100, 300
Argon	Ar, non-explosive inert gas, tasteless, odorless, colorless	131, 330 4,754 (liquid)
Carbon Dioxide	CO_2, non-explosive inert gas, tasteless, odorless, colorless (in large quantities is toxic)	20 lb, 50 lb
Helium	He, non-explosive inert gas, tasteless, odorless, colorless	221
Hydrogen	H_2, explosive gas, tasteless, odorless, colorless	191
Nitrogen	N_2, non-explosive inert gas, tasteless, odorless, colorless	20, 40, 80 113, 225
Oxygen	O_2, non-explosive gas, tasteless, odorless, colorless, supports combustion	20, 40, 80 122, 244 4,500 (liquid)

WELDING RODS – 36" LONG

Rod Size (Dia. in.)	Number of Rods per Pound			
	Aluminum	Brass	Cast Iron	Steel
$3/8$"	–	1.0	.25	1.0
$5/16$"	–	–	.50	1.33
$1/4$"	6.0	2.0	2.25	2.0
$3/16$"	9.0	3.0	5.50	3.5
$5/32$"	–	–	–	5.0
$1/8$"	23.0	7.0	–	8.0
$3/32$"	41.0	13.0	–	14.0
$1/16$"	91.0	29.0	–	31.0

PULLEY AND GEAR FORMULAS

For single reduction or increase of speed by means of belting where the speed at which each shaft should run is known, and one pulley is in place:

Multiply the diameter of the pulley that you have by the number of revolutions per minute its shaft makes; divide this product by the speed (in revolutions per minute) at which the second shaft should run. The result is the diameter of pulley to use.

Where both shafts with pulleys are in operation and the speed of one is known:

Multiply the speed of the shaft by diameter of its pulley and divide this product by diameter of pulley on the other shaft. The result is the speed of the second shaft.

Where a countershaft is used to obtain size of main driving or driven pulley, or speed of main driving or driven shaft, it is necessary to calculate, as above, between the known end of the transmission and the countershaft, then repeat this calculation between the countershaft and the unknown end.

A set of gears of the same pitch transmits speeds in proportion to the number of teeth they contain. Count the number of teeth in the gear wheel and use this quantity instead of the diameter of pulley, mentioned above, to obtain number of teeth cut in unknown gear, or speed of second shaft.

Formulas for Finding Pulley Sizes:

$$d = \frac{D \times S}{s'} \qquad D = \frac{d \times s'}{S}$$

d = diameter of driven pulley

D = diameter of driving pulley

s' = number of revolutions per minute of driven pulley

S = number of revolutions per minute of driving pulley

PULLEY AND GEAR FORMULAS *(cont.)*

Formulas for Finding Gear Sizes:

$$n = \frac{N \times S}{s'} \qquad N = \frac{n \times s'}{S}$$

n = number of teeth in pinion (driving gear)

N = number of teeth in gear (driven gear)

s' = number of revolutions per minute of gear

S = number of revolutions per minute of pinion

Formula to Determine Shaft Diameter:

$$\text{diameter of shaft in inches} = \sqrt[3]{\frac{K \times HP}{RPM}}$$

HP = the horsepower to be transmitted

RPM = speed of shaft

K = factor, which varies from 50 to 125 depending on type of shaft and distance between supporting bearings.

For line shaft having bearings 8 feet apart:

K = 90 for turned shafting

K = 70 for cold-rolled shafting

Formula to Determine Belt Length:

$$\text{length of belt} = \frac{3.14\,(D + d)}{2} + 2 \left(\sqrt{x^2 + \left(\frac{D - d}{2}\right)^2} \right)$$

D = diameter of large pulley

d = diameter of small pulley

X = distance between centers of shafting

STANDARD V BELT LENGTHS IN INCHES

A BELTS			B BELTS			C BELTS		
Standard Belt No.	Pitch Length	Outside Length	Standard Belt No.	Pitch Length	Outside Length	Standard Belt No.	Pitch Length	Outside Length
A26	27.3	28.0	B35	36.8	38.0	C51	53.9	55.0
A31	32.3	33.0	B38	39.8	41.0	C60	62.9	64.0
A35	36.3	37.0	B42	43.8	45.0	C68	70.9	81.0
A38	39.3	40.0	B46	47.8	49.0	C75	77.9	79.0
A42	43.3	44.0	B51	52.8	54.0	C81	83.9	85.0
A46	47.3	48.0	B55	56.8	58.0	C85	87.9	89.0
A51	52.3	53.0	B60	61.8	63.0	C90	92.9	94.0
A55	56.3	57.0	B68	69.8	71.0	C96	98.9	100.0
A60	61.3	62.0	B75	76.8	78.0	C105	107.9	109.0
A68	69.3	70.0	B81	82.8	84.0	C112	114.9	116.0
A75	76.3	77.0	B85	86.8	88.0	C120	122.9	124.0
A80	81.3	82.0	B90	91.8	93.0	C128	130.9	132.0
A85	86.3	87.0	B97	98.8	100.0	C136	138.9	140.0
A90	91.3	92.0	B105	106.8	108.0	C144	146.9	148.0
A96	97.3	98.0	B112	113.8	115.0	C158	160.9	162.0
A105	106.3	107.0	B120	121.8	123.0	C162	164.9	166.0
A112	113.3	114.0	B128	129.8	131.0	C173	175.9	177.0
A120	121.3	122.0	B136	137.8	139.0	C180	182.9	184.0
A128	129.3	130.0	B144	145.8	147.0	C195	197.9	199.0
			B158	159.8	161.0	C210	212.9	214.0
			B173	174.8	176.0	C240	240.9	242.0

D BELTS		
Standard Belt No.	Pitch Length	Outside Length

			B180	181.8	183.0	C270	270.9	272.0
D120	123.3	125.0	B195	196.8	198.0	C300	300.9	302.0
D128	131.3	133.0	B210	211.8	213.0	C360	360.9	362.0
D144	147.3	149.0	B240	240.3	241.5	C390	390.9	392.0
D158	161.3	163.0	B270	270.3	271.5	C420	420.9	422.0
D162	165.3	167.0	B300	300.3	301.5			

D173	176.3	178.0
D180	183.3	185.0

E BELTS					
Standard Belt No.	Pitch Length	Outside Length	Standard Belt No.	Pitch Length	Outside Length
D195	198.3	200.0						
D210	213.3	215.0	E180	184.5	187.5	E360	361.0	364.0
D240	240.8	242.0						
D270	270.8	272.5	E195	199.5	202.5	E390	391.0	394.0
D300	300.8	302.5						
D330	330.8	332.5	E210	214.5	217.5	E420	421.0	424.0
D360	360.8	362.5						
D390	390.8	392.5	E240	241.0	244.0	E480	481.0	484.0
D420	420.8	422.5						
D480	480.8	482.5	E270	271.0	274.0	E540	541.0	544.0
D540	540.8	542.5						
D600	600.8	602.5	E300	301.0	304.0	E600	601.0	604.0
			E330	331.0	334.0			

STANDARD V BELT LENGTHS IN INCHES *(cont.)*

3 V Belts		5 V Belts		8 V Belts	
3V250	25.0	5V500	50.0	8V1000	100.0
3V265	26.5	5V530	53.0	8V1060	106.0
3V280	28.0	5V560	56.0	8V1120	112.0
3V300	30.0	5V600	60.0	8V1180	118.0
3V315	31.5	5V630	63.0	8V1250	125.0
3V335	33.5	5V670	67.0	8V1320	132.0
3V355	35.5	5V710	71.0	8V1400	140.0
3V375	37.5	5V750	75.0	8V1500	150.0
3V400	40.0	5V800	80.0	8V1600	160.0
3V425	42.5	5V850	85.0	8V1700	170.0
3V450	45.0	5V900	90.0	8V1800	180.0
3V475	47.5	5V950	95.0	8V1900	190.0
3V500	50.0	5V1000	100.0	8V2000	200.0
3V530	53.0	5V1060	106.0	8V2120	212.0
3V560	56.0	5V1120	112.0	8V2240	224.0
3V600	60.0	5V1180	118.0	8V2360	236.0
3V630	63.0	5V1250	125.0	8V2500	250.0
3V670	67.0	5V1320	132.0	8V2650	265.0
3V710	71.0	5V1400	140.0	8V2800	280.0
3V750	75.0	5V1500	150.0	8V3000	300.0
3V800	80.0	5V1600	160.0	8V3150	315.0
3V850	85.0	5V1700	170.0	8V3350	335.0
3V900	90.0	5V1800	180.0	8V3550	355.0
3V950	95.0	5V1900	190.0	8V3750	375.0
3V1000	100.0	5V2000	200.0	8V4000	400.0
3V1060	106.0	5V2120	212.0	8V4250	425.0
3V1120	112.0	5V2240	224.0	8V4500	450.0
3V1180	118.0	5V2360	236.0	8V5000	500.0
3V1250	128.0	5V2500	250.0		
3V1320	132.0	5V2650	265.0		
3V1400	140.0	5V2800	280.0		
		5V3000	300.0		
		5V3150	315.0		
		5V3350	335.0		
		5V3550	355.0		

If a 60-inch "B" section belt is made 3/10 of an inch longer, it will be code marked 53 rather than 50. If made 3/10 shorter, it will be marked 47. Although both have the belt number B60 they cannot be used in a set because of the difference in length.

TYPICAL CODE MARKING

B60   Manufacturer's Name   50

Nominal
Size and Length

Length
Code Number

EXTENSION CORD SIZES FOR PORTABLE TOOLS

Cord Length (ft)	Full-Load Rating of Tool in Amperes at 115 Volts					
	0 to 2.0	2.1 to 3.4	3.5 to 5.0	5.1 to 7.0	7.1 to 12.0	12.1 to 16.0
	Wire Size (AWG)					
25	18	18	18	16	14	14
50	18	18	18	16	14	12
75	18	18	16	14	12	10
100	18	16	14	12	10	8
200	16	14	12	10	8	6
300	14	12	10	8	6	4
400	12	10	8	6	4	4
500	12	10	8	6	4	2
600	10	8	6	4	2	2
800	10	8	6	4	2	1
1,000	8	6	4	2	1	0

10-20

CHAPTER 11
Conversion Factors, Mathematics, and Units of Measurement

COMMONLY USED CONVERSION FACTORS		
Multiply	**By**	**To Obtain**
acres	43,560	square feet
acres	1.562×10^{-3}	square miles
acre-feet	43,560	cubic feet
amperes per sq cm	6.452	amperes per sq in
amperes per sq in	0.1550	amperes per sq cm
ampere-turns	1.257	gilberts
ampere-turns per cm	2.540	ampere-turns per in
ampere-turns per in	0.3937	ampere-turns per cm
atmospheres	76.0	cm of mercury
atmospheres	29.92	inches of mercury
atmospheres	33.90	feet of water
atmospheres	14.70	pounds per sq in
British thermal units	252.0	calories
British thermal units	778.2	foot-pounds
British thermal units	3.960×10^{-4}	horsepower-hours
British thermal units	0.2520	kilogram-calories
British thermal units	107.6	kilogram-meters
British thermal units	2.931×10^{-4}	kilowatt-hours
British thermal units	1,055	watt-seconds
Btu per hour	2.931×10^{-4}	kilowatts
Btu per minute	2.359×10^{-2}	horsepower
Btu per minute	1.759×10^{-2}	kilowatts
bushels	1.244	cubic feet
centimeters	0.3937	inches
circular mils	5.067×10^{-6}	square centimeters
circular mils	0.7854×10^{-6}	square inches

COMMONLY USED CONVERSION FACTORS *(cont.)*

Multiply	By	To Obtain
circular mils	0.7854	square mils
cords	128	cubic feet
cubic centimeters	6.102×10^{-6}	cubic inches
cubic feet	0.02832	cubic meters
cubic feet	7.481	gallons
cubic feet	28.32	liters
cubic inches	16.39	cubic centimeters
cubic meters	35.31	cubic feet
cubic meters	1.308	cubic yards
cubic yards	0.7646	cubic meters
degrees (angle)	0.01745	radians
dynes	2.248×10^{-6}	pounds
ergs	1	dyne-centimeters
ergs	7.37×10^{-6}	foot-pounds
ergs	10^{-7}	joules
farads	10^{6}	microfarads
fathoms	6	feet
feet	30.48	centimeters
feet of water	.08826	inches of mercury
feet of water	304.8	kg per square meter
feet of water	62.43	pounds per square ft
feet of water	0.4335	pounds per square in
foot-pounds	1.285×10^{-2}	British thermal units
foot-pounds	5.050×10^{-7}	horsepower-hours
foot-pounds	1.356	joules
foot-pounds	0.1383	kilogram-meters
foot-pounds	3.766×10^{-7}	kilowatt-hours
gallons	0.1337	cubic feet
gallons	231	cubic inches
gallons	3.785×10^{-3}	cubic meters
gallons	3.785	liters
gallons per minute	2.228×10^{-3}	cubic feet per sec.
gausses	6.452	lines per square in
gilberts	0.7958	ampere-turns
henries	10^{3}	millihenries
horsepower	42.41	Btu per min.
horsepower	2,544	Btu per hour

COMMONLY USED CONVERSION FACTORS *(cont.)*

Multiply	By	To Obtain
horsepower	550	foot-pounds per sec.
horsepower	33,000	foot-pounds per min.
horsepower	1.014	horsepower (metric)
horsepower	10.70	kg calories per min.
horsepower	0.7457	kilowatts
horsepower (boiler)	33,520	Btu per hour
horsepower-hours	2,544	British thermal units
horsepower-hours	1.98×10^6	foot-pounds
horsepower-hours	2.737×10^5	kilogram-meters
horsepower-hours	0.7457	kilowatt-hours
inches	2.540	centimeters
inches of mercury	1.133	feet of water
inches of mercury	70.73	pounds per square ft
inches of mercury	0.4912	pounds per square in
inches of water	25.40	kg per square meter
inches of water	0.5781	ounces per square in
inches of water	5.204	pounds per square ft
joules	9.478×10^{-4}	British thermal units
joules	0.2388	calories
joules	10^7	ergs
joules	0.7376	foot-pounds
joules	2.778×10^{-7}	kilowatt-hours
joules	0.1020	kilogram-meters
joules	1	watt-seconds
kilograms	2.205	pounds
kilogram-calories	3.968	British thermal units
kilogram meters	7.233	foot-pounds
kg per square meter	3.281×10^{-3}	feet of water
kg per square meter	0.2048	pounds per square ft
kg per square meter	1.422×10^{-3}	pounds per square in
kilolines	10^3	maxwells
kilometers	3.281	feet
kilometers	0.6214	miles
kilowatts	56.87	Btu per min.
kilowatts	737.6	foot-pounds per sec.
kilowatts	1.341	horsepower
kilowatts-hours	3409.5	British thermal units

COMMONLY USED CONVERSION FACTORS *(cont.)*

Multiply	By	To Obtain
kilowatts-hours	2.655×10^6	foot-pounds
knots	1.152	miles
liters	0.03531	cubic feet
liters	61.02	cubic inches
liters	0.2642	gallons
log N_e or in N	0.4343	$\log_{10}$ N
log N	2.303	$\log_e$ N or in N
lumens per square ft.	1	footcandles
maxwells	10^{-3}	kilolines
megalines	10^6	maxwells
megaohms	10^6	ohms
meters	3.281	feet
meters	39.37	inches
meter-kilograms	7.233	pound-feet
microfarads	10^{-6}	farads
microhms	10^{-6}	ohms
microhms per cm cube	0.3937	microhms per in cube
microhms per cm cube	6.015	ohms per mil foot
miles	5,280	feet
miles	1.609	kilometers
miner's inches	1.5	cubic feet per min.
ohms	10^{-6}	megohms
ohms	10^6	microhms
ohms per mil foot	0.1662	microhms per cm cube
ohms per mil foot	0.06524	microhms per in cube
poundals	0.03108	pounds
pounds	32.17	poundals
pound-feet	0.1383	meter-kilograms
pounds of water	0.01602	cubic feet
pounds of water	0.1198	gallons
pounds per cubic foot	16.02	kg per cubic meter
pounds per cubic foot	5.787×10^{-4}	pounds per cubic in
pounds per cubic inch	27.68	grams per cubic cm
pounds per cubic inch	2.768×10^{-4}	kg per cubic meter
pounds per cubic inch	1.728	pounds per cubic ft
pounds per square foot	0.01602	feet of water
pounds per square foot	4.882	kg per square meter

COMMONLY USED CONVERSION FACTORS *(cont.)*

Multiply	By	To Obtain
pounds per square foot	6.944×10^{-3}	pounds per sq in
pounds per square inch	2.307	feet of water
pounds per square inch	2.036	inches of mercury
pounds per square inch	703.1	kg per square meter
radians	57.30	degrees
square centimeters	1.973×10^{5}	circular mils
square feet	2.296×10^{-5}	acres
square feet	0.09290	square meters
square inches	1.273×10^{6}	circular mils
square inches	6.452	square centimeters
square kilometers	0.3861	square miles
square meters	10.76	square feet
square miles	640	acres
square miles	2.590	square kilometers
square millimeters	1.973×10^{3}	circular mils
square mils	1.273	circular mils
tons (long)	2,240	pounds
tons (metric)	2,205	pounds
tons (short)	2,000	pounds
watts	0.05686	Btu per minute
watts	10^{7}	ergs per sec.
watts	44.26	foot-pounds per min.
watts	1.341×10^{-3}	horsepower
watts	14.34	calories per min.
watts-hours	3.412	british thermal units
watts-hours	2,655	footpounds
watts-hours	1.341×10^{-3}	horsepower-hours
watts-hours	0.8605	kilogram-calories
watts-hours	376.1	kilogram-meters
webers	10^{8}	maxwells

METRIC TO TRADE SIZE

METRIC DESIGNATOR (mm)

12	16	21	27	35	41	53	63	78	91	103	129	155
⅜	½	¾	1	1¼	1½	2	2½	3	3½	4	5	6

TRADE SIZE (in)

DECIMAL EQUIVALENTS OF FRACTIONS

8ths	32nds	64ths	64ths
1/8 = .125	1/32 = .03125	1/64 = .015625	33/64 = .515625
1/4 = .250	3/32 = .09375	3/64 = .046875	35/64 = .546875
3/8 = .375	5/32 = .15625	5/64 = .078125	37/64 = .57812
1/2 = .500	7/32 = .21875	7/64 = .109375	39/64 = .609375
5/8 = .625	9/32 = .28125	9/64 = .140625	41/64 = .640625
3/4 = .750	11/32 = .34375	11/64 = .171875	43/64 = .671875
7/8 = .875	13/32 = .40625	13/64 = .203128	45/64 = .703125
16ths	15/32 = .46875	15/64 = .234375	47/64 = .734375
1/16 = .0625	17/32 = .53125	17/64 = .265625	49/64 = .765625
3/16 = .1875	19/32 = .59375	19/64 = .296875	51/64 = 3796875
5/16 = .3125	21/32 = .65625	21/64 = .328125	53/64 = .828125
7/16 = .4375	23/32 = .71875	23/64 = .359375	55/64 = .859375
9/16 = .5625	25/32 = .78125	25/64 = .390625	57/64 = .890625
11/16 = .6875	27/32 = .84375	27/64 = .421875	59/64 = .921875
13/16 = .8125	29/32 = .90625	29/64 = .453125	61/64 = .953125
15/16 = .9375	31/32 = .96875	31/64 = .484375	63/64 = .984375

11-6

MILLIMETER AND DECIMAL INCH EQUIVALENTS

mm in	mm in	mm in	mm in	mm in
$\frac{1}{50}$ = .00079	$\frac{30}{50}$ = .02362	11 = .43307	41 = 1.61417	71 = 2.79527
$\frac{2}{50}$ = .00157	$\frac{31}{50}$ = .02441	12 = .47244	42 = 1.65354	
$\frac{3}{50}$ = .00236	$\frac{32}{50}$ = .02520	13 = .51181	43 = 1.69291	72 = 2.83464
$\frac{4}{50}$ = .00315	$\frac{33}{50}$ = .02598	14 = .55118	44 = 1.73228	73 = 2.87401
	$\frac{34}{50}$ = .02677			74 = 2.91338
$\frac{5}{50}$ = .00394		15 = .59055	45 = 1.77165	75 = 2.95275
$\frac{6}{50}$ = .00472	$\frac{35}{50}$ = .02756	16 = .62992	46 = 1.81102	76 = 2.99212
$\frac{7}{50}$ = .00551	$\frac{36}{50}$ = .02835	17 = .66929	47 = 1.85039	
$\frac{8}{50}$ = .00630	$\frac{37}{50}$ = .02913	18 = .70866	48 = 1.88976	77 = 3.03149
$\frac{9}{50}$ = .00709	$\frac{38}{50}$ = .02992	19 = .74803	49 = 1.92913	78 = 3.07086
	$\frac{39}{50}$ = .03071			79 = 3.11023
		20 = .78740	50 = 1.96850	80 = 3.14960
$\frac{10}{50}$ = .00787	$\frac{40}{50}$ = .03150	21 = .82677	51 = 2.00787	81 = 3.18897
$\frac{11}{50}$ = .00866	$\frac{41}{50}$ = .03228	22 = .86614	52 = 2.04724	
$\frac{12}{50}$ = .00945	$\frac{42}{50}$ = .03307	23 = .90551	53 = 2.08661	82 = 3.22834
$\frac{13}{50}$ = .01024	$\frac{43}{50}$ = .03386	24 = .94488	54 = 2.12598	83 = 3.26771
$\frac{14}{50}$ = .01102	$\frac{44}{50}$ = .03465			84 = 3.30708
		25 = .98425	55 = 2.16535	85 = 3.34645
$\frac{15}{50}$ = .01181	$\frac{45}{50}$ = .03543	26 = 1.02362	56 = 2.20472	86 = 3.38582
$\frac{16}{50}$ = .01260	$\frac{46}{50}$ = .03622	27 = 1.06299	57 = 2.24409	
$\frac{17}{50}$ = .01339	$\frac{47}{50}$ = .03701	28 = 1.10236	58 = 2.28346	87 = 3.42519
$\frac{18}{50}$ = .01417	$\frac{48}{50}$ = .03780	29 = 1.14173	59 = 2.32283	88 = 3.46456
$\frac{19}{50}$ = .01496	$\frac{49}{50}$ = .03858			89 = 3.50393
		30 = 1.18110	60 = 2.36220	90 = 3.54330
$\frac{20}{50}$ = .01575	1 = .03937	31 = 1.22047	61 = 2.40157	91 = 3.58267
$\frac{21}{50}$ = .01654	2 = .07874	32 = 1.25984	62 = 2.44094	
$\frac{22}{50}$ = .01732	3 = .11811	33 = 1.29921	63 = 2.48031	92 = 3.62204
$\frac{23}{50}$ = .01811	4 = .15748	34 = 1.33858		93 = 3.66141
$\frac{24}{50}$ = .01890			64 = 2.51968	94 = 3.70078
	5 = .19685	35 = 1.37795	65 = 2.55905	95 = 3.74015
	6 = .23622	36 = 1.41732	66 = 2.59842	96 = 3.77952
$\frac{25}{50}$ = .01969	7 = .27559	37 = 1.45669		97 = 3.81889
$\frac{26}{50}$ = .02047	8 = .31496	38 = 1.49606	67 = 2.63779	98 = 3.85826
$\frac{27}{50}$ = .02126	9 = .35433	39 = 1.53543	68 = 2.67716	99 = 3.89763
$\frac{28}{50}$ = .02205			69 = 2.71653	
$\frac{29}{50}$ = .02283	10 = .39370	40 = 1.57480	70 = 2.75590	100 = 3.93700

METRIC MEASUREMENTS

Prefixes

mega	=	1,000,000	deci	=	0.1
kilo	=	1,000	centi	=	0.01
hecto	=	100	milli	=	0.001
deka	=	10	micro	=	0.000001

Linear Measures

(1 meter = 39.37 inches)

1 centimeter	=	10 millimeters	=	0.3937 in
1 decimeter	=	10 centimeters	=	3.937 in
1 meter	=	10 decimeters	=	1.0936 yd
1 dekameter	=	10 meters	=	10.9361 yd
1 hectometer	=	10 dekameters	=	109.3614 yd
1 kilometer	=	10 hectometers	=	0.6213 mi

Square Measures

(1 square meter = 1549.997 square inches)

1 sq centimeter	= 100 sq millimeters	= 0.155 sq in
1 sq decimeter	= 100 sq centimeters	= 15.55 sq in
1 sq meter	= 100 sq decimeters	= 10.764 sq ft
1 sq dekameter	= 100 sq meters	= 119.6 sq yd
1 sq hectometer	= 100 sq dekameters	
1 sq kilometer	= 100 sq hectometers	

(1 are = 100 sq meters)

1 centiare	=	10 milliares	=	10.764 sq ft
1 deciare	=	10 centiares	=	11.96 sq yd
1 are	=	10 deciares	=	119.6 sq yd
1 dekare	=	10 ares	=	0.247 acres
1 hektare	=	10 dekares	=	2.471 acres
1 sq kilometer	=	100 hektares	=	0.386 sq mi

METRIC MEASUREMENTS *(cont.)*

Cubic Measures

(1 meter = 39.37 inches)

1 cu centimeter	= 1000 cu millimeters	= 0.061 cu in
1 cu decimeter	= 1000 cu centimeters	= 61.023 cu in
1 cu meter	= 1000 cu decimeters	= 35.314 cu ft
1 cu meter	= 1 stere	= 1.3079 cu yd
1 cu centimeter		= 1 g
1000 cu centimeters (1 liter)		= 1 kg
1 cu meter (1,000 liters)		= 1 metric ton

Weight Measures

(1 gram = 0.0352 ounces)

1 centigram	= 10 milligrams	=	0.154 grains
1 decigram	= 10 centigrams	=	1.543 grains
1 gram	= 10 decigrams	=	15.432 grains
1 dekagram	= 10 grams	=	5.643 drams
1 hectogram	= 10 dekagrams	=	3.527 ounces
1 kilogram	= 10 hectograms	=	2.2046 pounds
1 myriagram	= 10 kilograms	=	22.0462 pounds
1 quintal	= 10 myriagrams	=	220.4622 pounds
1 metric ton	= 10 quintal	=	2,204.622 pounds

Capacity Measures

(1 liter = 1.0567 liquid quarts)

1 centiliter	= 10 milliliters	=	0.338 fluid ounces
1 deciliter	= 10 centiliters	=	3.38 fluid ounces
1 liter	= 10 deciliters	=	33.8 fluid ounces
1 dekaliter	= 10 liters	=	0.284 bushel
1 hectoliter	= 10 dekaliters	=	2.84 bushels
1 kiloliter	= 10 hectoliters	=	264.2 gallons

U.S. MEASUREMENTS

Linear Measures

			1	inch	=	2.540 centimeters
12	inches	=	1	foot	=	3.048 decimeters
3	feet	=	1	yard	=	9.144 decimeters
5.5	yards	=	1	rod	=	5.029 meters
40	rods	=	1	furlong	=	2.018 hectometers
8	furlongs	=	1	mile	=	1.609 kilometers
1	stature mile	=	5,280	feet		
1	span	=	9	inches		
1	chain	=	22	yards		
1	link	=	7.92	inches		
1	fathom	=	6	feet		
1	furlong	=	10	chains		
1	cable	=	608	feet		

Square Measures

144	square inches	=	1 square foot
9	square feet	=	1 square yard
30.25	square yards	=	1 square rod
40	rods	=	1 rood
4	roods	=	1 acre
640	acres	=	1 square mile
1	square mile	=	1 section
36	sections	=	1 township

Cubic or Solid Measures

1 cu foot	=	1,728 cu inches
1 cu yard	=	27 cu feet
1 cu foot	=	7.48 gallons
1 gallon (water)	=	8.34 lb
1 gallon (u.s.)	=	231 cu inches of water
1 gallon (imperial)	=	277.25 cu inches of water

U.S. MEASUREMENTS *(cont.)*

Liquid Measures

1 u.s. pint	=	16 fl. ounces
1 standard cup	=	8 fl. ounces
1 tablespoon	=	0.5 fl. ounces
1 teaspoon	=	0.16 fl. ounces
1 pint	=	2 cups
1 quart	=	2 pints
1 gallon	=	4 quarts
1 barrel	=	42 gallons (crude oil)

Dry Measures

1 bushel	=	8 gallons

Weight (Mass) Measures

Avoirdupois Weight:

1 ounce	=	16 drams
1 pound	=	16 ounces
1 hundredweight	=	100 pounds
1 ton	=	2,000 pounds

Troy Weight:

1 carat	=	3.17 grains
1 pennyweight	=	20 grains
1 ounce	=	20 pennyweights
1 pound	=	12 ounces
1 long hundred-weight	=	112 pounds
1 long ton	=	2,240 pounds

Apothecaries Weight:

1 scruple	=	20 grains	=	1.296 grams
1 dram	=	3 scruples	=	3.888 grams
1 ounce	=	8 drams	=	31.10 grams
1 pound	=	12 ounces	=	373.24 grams

MATHEMATICS

Formulas

Formulas are statements of truth. If we say that $X = 1 + 2$, we can then manipulate the formula to say $X - 2 = 1$, or, $X - 1 = 2$. This is called *transposing* a formula. (Obviously, $X = 3$.)

Add or Subtract from Both Sides

Formulas are balanced statements of equality. Changing one side and not the other puts it out of balance, but changing both sides equally is fine.

Subtract 5 from each side: $x = y + 5$

$x - 5 = y + 5 - 5$ Since $+5 - 5 = 0$, $x - 5 = y$

$a = b - 5$

Add 5 to each side:

$a + 5 = b - 5 + 5$ $a + 5 = b$

Multiplying or Dividing Both Sides

You can multiply or divide, so long as you do it equally to both sides.

$x = y \div 4$ Multiply both sides by 4:

$4x = (y \div 4) \times 4$,   Since $\div 4$ and $\times 4$ cancel, $4x = y$

$a = 4b$ Divide each side by 4: $a \div 4 = 4b \div 4$

since $\times b$ and $\div b$ cancel, $a \div 4 = b$

Performing Operations

You can perform other operations to each side of a formula without throwing it out of balance.

For example, we can square each side of an equation without ruining its effectiveness as a true statement. Here is an example of this:

$c = \sqrt{d}$ Squaring both sides: $c^2 = d$

Square Roots

A *square* is any number multiplied by itself.

We note a square with a superscript. "5 squared" $= 5^2$.

In addition to squares (also called "to the second power"), we can raise numbers to higher powers as well. "5 to the third power" is written 5^3, and would equal $5 \times 5 \times 5$, or 125.

A square root is the number that multiplied by itself would equal the number in question. The $\sqrt{}$ symbol denotes square root.

MATHEMATICS *(cont.)*

Fractions

Fractions are statements of division. The top portion of the fraction is called the numerator and the bottom portion is called the denominator.

$$\frac{\text{numerator}}{\text{denominator}}$$

For this: $\frac{2}{3}$,

we can say "2 over 3" or "2 divided by 3."

$$\frac{2}{3} = 2 \div 3 = .67$$

Adding and Subtracting Fractions

Adding fractions that have the same denominator is simple. The denominators stay the same, and we simply add or subtract the numerators.

$$\frac{1}{8} + \frac{4}{8} + \frac{5}{8} = \frac{10}{8} \text{ or } 1\frac{2}{8}$$

$$\frac{7}{16} - \frac{2}{16} = \frac{5}{16}$$

If the denominators are not equal, we first create identical denominators.

$$\frac{1}{3} + \frac{1}{4}$$

$$\frac{1}{3} = \frac{4}{12} \qquad\qquad \frac{1}{4} = \frac{3}{12}$$

$$\frac{4}{12} + \frac{3}{12} = \frac{7}{12} \text{ , thus } \frac{1}{3} + \frac{1}{4} = \frac{7}{12}$$

MATHEMATICS (cont.)

Multiplying Fractions
To multiply fractions you multiply the numerator by the numerator and the denominator by the denominator:

$$\frac{2}{3} \times \frac{1}{2} = \frac{2}{6} = \frac{1}{3}$$

Dividing Fractions
Dividing fractions is exactly the same as multiplication, except you invert the second fraction before multiplying, as shown here:

$$\frac{2}{3} \div \frac{1}{2} = \frac{2}{3} \times \frac{2}{1} = \frac{4}{3}$$

Decimals
Decimals are actually a type of fraction that always have a multiple of ten in their denominator.

$$\frac{42}{10} = 4.2$$

$$\frac{5}{100} = .05$$

$$\frac{28}{100} = .28$$

When multiplying numbers by factors of 10, the decimal point is moved to the right.

$51.45 \times 10 = 514.5$
$51.45 \times 100 = 5,145.0$
$51.45 \times 1000 = 51,450.0$

When dividing by factors of 10, the decimal point is moved to the left.

$62.56 \div 10 = 6.256$
$62.56 \div 100 = .6256$
$62.56 \div 1000 = .06256$

Scientific Notation
Scientific notation (also referred to as *exponents*) is a compact method of expressing very large or very small numbers. Numbers can be expressed as 1 through 10 multiplied by a power of ten. Values above 10 have positive exponents, values less than 1 have negative exponents.

$420,000 = 4.2 \times 10^5$
$.00214 = 2.14 \times 10^{-3}$

MATHEMATICS *(cont.)*

The Pythagorean Theorem

Describes the relationship of the sides of a **right triangle.** A right triangle always has one 90 degree angle. The longest side of a triangle is called the **hypotenuse.**

The Pythagorean Theorem states:

$$C^2 = A^2 + B^2, \quad \text{or,} \quad C = \sqrt{A^2 + B^2}$$

This formula is true for every right triangle, and it is very useful. A simple right triangle is one with side measurements of 3, 4, and 5.

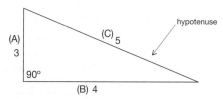

Trigonometric Functions

The primary functions are sine, cosine, and tangent, which state the relationship between the sides of a right triangle.

$$\text{sine} = \frac{\text{opposite}}{\text{hypotenuse}}$$

$$\text{cosine} = \frac{\text{adjacent}}{\text{hypotenuse}}$$

$$\text{tangent} = \frac{\text{opposite}}{\text{adjacent}}$$

COMMONLY USED GEOMETRICAL RELATIONSHIPS

Diameter of a circle × 3.1416 = Circumference

Radius of a circle × 6.283185 = Circumference

Square of the radius of a circle × 3.1416 = Area

Square of the diameter of a circle × 0.7854 = Area

Square of the circumference of a circle × 0.07958 = Area

Half the circumference of a circle × half its diameter = Area

Circumference of a circle × 0.159155 = Radius

Square root of the area of a circle × 0.56419 = Radius

Circumference of a circle × 0.31831 = Diameter

Square root of the area of a circle × 1.12838 = Diameter

Diameter of a circle × 0.866 = Side of an inscribed equilateral triangle

Diameter of a circle × 0.7071 = Side of an inscribed square

Circumference of a circle × 0.225 = Side of an inscribed square

Circumference of a circle × 0.282 = Side of an equal square

Diameter of a circle × 0.8862 = Side of an equal square

Base of a triangle × one-half the altitude = Area

Multiplying both diameters and .7854 together = Area of an ellipse

Surface of a sphere × one-sixth of its diameter = Volume

Circumference of a sphere × its diameter = Surface

Square of the diameter of a sphere × 3.1416 = Surface

Square of the circumference of a sphere × 0.3183 = Surface

Cube of the diameter of a sphere × 0.5236 = Volume

Cube of the circumference of a sphere × 0.016887 = Volume

Radius of a sphere × 1.1547 = Side of an inscribed cube

Diameter of a sphere divided by $\sqrt{3}$ = Side of an inscribed cube

Area of its base × one-third of its altitude = Volume of a cone or
pyramid whether round, square, or triangular

Area of one of its sides × 6 = Surface of the cube

Altitude of trapezoid × one-half the sum of its parallel sides = Area

COMMON ENGINEERING UNITS AND THEIR RELATIONSHIP

Quantity	SI Metric Units/Symbols	Customary Units	Relationship of Units
Acceleration	meters per second squared (m/s^2)	feet per second squared (ft/s^2)	m/s^2 = ft/s^2 × 3.281
Area	square meter (m^2) square millimeter (mm^2)	square foot (ft^2) square inch (in^2)	m^2 = ft^2 × 10.764 mm^2 = in^2 × 0.00155
Density	kilograms per cubic meter (kg/m^3) grams per cubic centimeter (g/cm^3)	pounds per cubic foot (lb/ft^3) pounds per cubic inch (lb/in^3)	kg/m^3 = lb/ft^2 × 16.02 g/cm^3 = lb/in^3 × 0.036
Work	Joule (J)	foot pound force (ft lbf or ft lb)	J = ft lbf × 1.356
Heat	Joule (J)	British thermal unit (Btu) Calorie (Cal)	J = Btu × 1.055 J = cal × 4.187
Energy	kilowatt (kW)	Horsepower (HP)	kW = HP × 0.7457
Force	Newton (N) Newton (N)	Pound-force (lbf, lb · f, or lb) kilogram-force (kgf, kg · f, or kp)	N = lbf × 4.448 $N = \dfrac{kgf}{9.807}$
Length	meter (m) millimeter (mm)	foot (ft) inch (in)	m = ft × 3.281 $mm = \dfrac{in}{25.4}$
Mass	kilogram (kg) gram (g)	pound (lb) ounce (oz)	kg = lb × 2.2 $g = \dfrac{oz}{28.35}$
Stress	Pascal = Newton per second (Pa = N/s)	pounds per square inch (lb/in^2 or psi)	Pa = lb/in^2 × 6,895
Temperature	degree Celsius (°C)	degree Fahrenheit (°F)	$°C = \dfrac{°F - 32}{1.8}$
Torque	Newton meter (N · m)	foot-pound (ft lb) inch-pound (in lb)	N · m = ft lbf × 1.356 N · m = in lbf × 0.113
Volume	cubic meter (m^3) cubic centimeter (cm^3)	cubic foot (ft^3) cubic inch (in^3)	m^3 = ft^3 × 35.314 $cm^3 = \dfrac{in^3}{16.387}$

CONVERSION TABLE FOR TEMPERATURE – °F/°C

°F	°C	°F	°C	°F	°C	°F	°C	°F	°C
-459.4	-273	-22.0	-30	35.6	2	93.2	34	150.8	66
-418.0	-250	-18.4	-28	39.2	4	96	36	154.4	68
-328.0	-200	-14.8	-26	42.8	6	100.4	38	158.0	70
-238.0	-150	-11.2	-24	46.4	8	104.0	40	161.6	72
-193.0	-125	-7.6	-22	50.0	10	107.6	42	165.2	74
-148.0	-100	-4.0	-20	53.6	12	111.2	44	168.8	76
-130.0	-90	-0.4	-18	57.2	14	114.8	46	172.4	78
-112.0	-80	3.2	-16	60.8	16	118.4	48	176.0	80
-94.0	-70	6.8	-14	64.4	18	122.0	50	179.6	82
-76.0	-60	10.4	-12	68.0	20	125.6	52	183.2	84
-58.0	-50	14.0	-10	71.6	22	129.2	54	186.8	86
-40.0	-40	17.6	-8	75.2	24	132.8	56	190.4	88
-36.4	-38	21.2	-6	78.8	26	136.4	58	194.0	90
-32.8	-36	24.8	-4	82.4	28	140.0	60	197.6	92
-29.2	-34	28.4	-2	86.0	30	143.6	62	201.2	94
-25.6	-32	32.0	0	89.6	32	147.2	64	204.8	96

1 degree F is 1/180 of the difference between the temperature of melting ice and boiling water.
1 degree C is 1/100 of the difference between the temperature of melting ice and boiling water.

Absolute Zero = 273.16°C = -459.69°F

11-18

CONVERSION TABLE FOR TEMPERATURE – °F/°C (cont.)

°F	°C	°F	°C	°F	°C	°F	°C	°F	°C
208.4	98	347	175	590	310	1,004	540	6,332	3,500
212	100	356	180	608	320	1,040	560	7,232	4,000
221	105	365	185	626	330	1,076	580	8,132	4,500
230	110	374	190	644	340	1,112	600	9,032	5,000
239	115	383	195	662	350	1,202	650	9,932	5,500
248	120	392	200	680	360	1,292	700	10,832	6,000
257	125	410	210	698	370	1,382	750	11,732	6,500
266	130	428	220	716	380	1,472	800	12,632	7,000
275	135	446	230	734	390	1,562	850	13,532	7,500
284	140	464	240	752	400	1,652	900	14,432	8,000
293	145	482	250	788	420	1,742	950	15,332	8,500
302	150	500	260	824	440	1,832	1,000	16,232	9,000
311	155	518	270	860	460	2,732	1,500	17,132	9,500
320	160	536	280	896	480	3,632	2,000	18,032	10,000
329	165	554	290	932	500	4,532	2,500		
338	170	572	300	968	520	5,432	3,000		

1 degree F is 1/180 of the difference between the temperature of melting ice and boiling water.
1 degree C is 1/100 of the difference between the temperature of melting ice and boiling water.

Absolute Zero = 273.16°C = –459.69°F

ELECTRICAL PREFIXES

Prefixes
Prefixes are used to avoid long expressions of units that are smaller and larger than the base unit. See Common Prefixes. For example, sentences 1 and 2 do not use prefixes. Sentences 3 and 4 use prefixes.
1. A solid-state device draws 0.000001 amperes (A).
2. A generator produces 100,000 watts (W).
3. A solid-state device draws 1 microampere (μA).
4. A generator produces 100 kilowatts (kW).

Converting Units
To convert between different units, the decimal point is moved to the left or right, depending on the unit. See Conversion Table. For example, an electronic circuit has a current flow of .000001 A. The current value is converted to simplest terms by moving the decimal point six places to the right to obtain 1.0μA (from Conversion Table).

$$.000001. \ A = 1.0 \ \mu A$$

Move decimal point
6 places to right

Common Electrical Quantities
Abbreviations are used to simplify the expression of common electrical quantities.
See Common Electrical Quantities. For example, milliwatt is abbreviated mW, kilovolt is abbreviated kV, and ampere is abbreviated A.

COMMON PREFIXES

Symbol	Prefix	Equivalent
G	giga	1,000,000,000
M	mega	1,000,000
k	kilo	1000
base unit	—	1
m	milli	.001
u	micro	.000001
n	nano	.000000001

COMMON ELECTRICAL QUANTITIES

Variable	Name	Unit of Measure and Abbreviation
E	voltage	volt - V
I	current	ampere - A
R	resistance	ohm - Ω
P	power	watt - W
P	power (apparent)	volt-amp - VA
C	capacitance	farad - F
L	inductance	henry - H
Z	impedance	ohm - Ω
G	conductance	siemens - S
f	frequency	hertz - Hz
T	period	second - s

CONVERSION TABLE

Initial Units	Final Units						
	giga	mega	kilo	base unit	milli	micro	nano
giga	—	3R	6R	9R	12R	15R	18R
mega	3L	—	3R	6R	9R	12R	15R
kilo	6L	3L	—	3R	6R	9R	12R
base unit	9L	6L	3L	—	3R	6R	9R
milli	12L	9L	6L	3L	—	3R	6R
micro	15L	12L	9L	6L	3L	—	3R
nano	18L	15L	12L	9L	6L	3L	—

CHAPTER 12
Symbols

SWITCH OUTLETS	
S	Single-pole switch
S_2	Double-pole switch
S_3	Three-way switch
S_4	Four-way switch
S_K	Key-operated switch
S_P	Switch and pilot lamp
S_{wp}	Weather-proof switch
S_F	Fused switch
⊖ S	Switch and single receptacle
⊖ S	Switch and duplex receptacle

SWITCH OUTLETS *(cont.)*

S_D	Door switch
S_T	Time switch
S_{CB}	Circuit-breaker switch
S_{MC}	Momentary contact switch or pushbutton for other than signaling system

LIGHTING OUTLETS

CEILING	WALL	
◯	─◯	Surface or pendant incandescent, mercury vapor, or similar lamp fixture
(R)	─(R)	Recessed incandescent, mercury vapor, or similar lamp fixture
[○]		Surface or pendant individual fluorescent fixture
[○ R]		Recessed individual fluorescent fixture

⟨O──────────────⟩	Surface or pendant continuous-row fluorescent fixture
⟨O R───────────⟩	Recessed continuous-row fluorescent fixture*
├────┼────┼────┤	Bare-lamp fluorescent strip**

CEILING	WALL	
(X)	—(X)	Surface or pendant exit light
(XR)	—(XR)	Recessed exit light
(B)	—(B)	Blanked outlet
(J)	—(J)	Junction box

* In the case of combination continuous-row fluorescent and incandescent spotlights, use combinations of the above standard symbols.

** In the case of a continuous-row bare-lamp fluorescent strip above an area-wide diffusion means, show each fixture run using the standard symbol; indicate area of diffusing means and type of light shading and/or drawing notation.

RECEPTACLE OUTLETS

Single receptacle outlet

Duplex receptacle outlet

Triplex receptacle outlet

Quadruplex receptacle outlet

Duplex receptacle
outlet-split wired

Single special-purpose
receptacle outlet*

Duplex special-purpose
receptacle outlet*

Range outlet

Unless noted, assume every receptacle will be grounded and will have a separate grounding contact.

* Use a numeral or letter, either within the symbol or as a subscript alongside the symbol keyed to explanation in the drawing list of symbols, to indicate type of receptacle or usage.

RECEPTACLE OUTLETS *(cont.)*

 Special-purpose connection or provision for connection. Use subscript letters to indicate function (DW—dishwasher; CD—clothes dryer, etc.)

 Multi-outlet assembly. Extend arrows to limit of installation. Use appropriate symbol to indicate type of outlet. Also indicate spacing of outlets as × inches.

 Pushbutton

 Buzzer

 Bell

 Chime

 Annunciator

Electric door opener

 Telephone outlet

Television outlet

Unless noted, assume every receptacle will be grounded and will have a separate grounding contact.

* Use numeral or letter, either within the symbol or as a subscript alongside the symbol keyed to explanation in the drawing list of symbols, to indicate type of receptacle or usage.

RECEPTACLE OUTLETS *(cont.)*

(C)	Clock hanger receptacle
(F)	Fan hanger receptacle
⊖	Floor outlet, duplex receptacle
◬	Floor outlet*, special-purpose
◁	Floor outlet, telephone

GENERAL

(M)	Electric motor
(G)	Electric generator
⟩⟩⟩	Transformer
(WH)	Watt-hour meter

Unless noted, assume every receptacle will be grounded and will have a separate grounding contact.

* Use numeral or letter, either within the symbol or as a subscript alongside the symbol keyed to explanation in the drawing list of symbols, to indicate type of receptacle or usage.

Circuit breaker

Fusible element

Single-throw knife switch

Double-throw knife switch

Ground

Battery

M	Manhole
H	Handhole
TM	Transformer manhole or vault
TP	Transformer pad
–––– –– –– –– ––––	Underground direct burial cable. Indicate type, size, and number of conductors by notation or schedule.
	Underground duct line. Indicate type, size, and number of ducts by cross-section identification of each run by notation or schedule. Indicate type, size, and number of conductors by notation or schedule.
	Streetlight standard feed from underground circuit

PANELBOARDS, SWITCHBOARDS, AND RELATED EQUIPMENT

Flush-mounted panelboard or cabinet

Surface-mounted panelboard or cabinet

Switchboard, power control center, unit substations — should be drawn to scale

Motor or other power controller

Externally-operated disconnection switch

Combination controller and disconnection means

CIRCUITING AND WIRING

Wiring concealed in ceiling or wall

Wiring concealed in floor

Wiring exposed

Branch-circuit home run to panelboard. Number of arrows indicates number of circuits. Any circuit without further identification indicates two-wire circuit. For a greater number of wires, indicate with cross lines.

CIRCUITING AND WIRING *(cont.)*

4 wires

Unless indicated otherwise, the wire size of the circuit is the minimum size required by the specification.

Wiring turned up

Wiring turned down

MOTOR CONTROL

Momentary-contact switch, normally open

Momentary-contact switch, normally closed

Inductor, Iron-core

Armature

Crossed wires, not connected

Crossed wires, connected

Fuse

Limit switch, normally open

Limit switch, normally closed

Footswitch, normally open

Footswitch, normally closed

Vacuum switch, normally open

Vacuum switch, normally closed

Liquid-level switch, normally open

Liquid-level switch, normally closed

MOTOR CONTROL *(cont.)*

Temperature-actuated switch, normally open

Temperature-actuated switch, normally closed

Flow switch, normally open

Flow switch, normally closed

Magnetic blowout coil

Manually operated 3-pole contactor

Thermocouple

Diode (rectifier)

Capacitor

MOTOR CONTROL (cont.)

—+⊢—	Adjustable capacitor
—⋀⋀—	Resistor
—⋀⋀—	Tapped resistor
—⋀⋀⋀—	Variable resistor
——o——	Wiring terminal
	Full-wave rectifier
-------------------	Mechanical interlock
----------+----------	Mechanical connection

SWITCH

Switch symbols are constructed of basic symbols for mechanical connections, contacts, etc. and normally a switch is represented in the de-energized position for switches having two or more positions where no operating force is applied. When actuated by a mechanical force, the functioning point is described by a clarifying note. Where switch symbols are in closed position, the terminals should be included for clarity.

Single throw, general

Double throw, general

2-pole double throw switch with terminals shown

Knife switch, general

PUSHBUTTON, MOMENTARY, OR SPRING RETURN

Circuit closing (make)

Circuit opening (break)

Two-circuit

PUSHBUTTON, MAINTAINED, OR NO-SPRING RETURN

Two-circuit

12-14

CHAPTER 13
Glossary

10Base2: 10 Mbps, baseband, in 185 meter segments. The IEEE 802.3 substandard for ThinWire, coaxial, Ethernet.

10BaseT: 10 Mbps, baseband, over twisted pair. The IEEE 802.3 substandard for unshielded twisted pair Ethernet.

110 Type Block: A connecting block that terminates 100 to 300 pairs of wires. Used for telephone and communications wiring.

66-Type Block: A type of wire-connecting block that is used for twisted pair cabling cross connections. It holds 25 pairs in up to four columns.

AC Current: Electrical current that reverses direction at regular intervals (cycles) due to a change in voltage occurring at the same frequency.

AC Line Filter: Absorbs electrical interference from motors, etc.

AC Sine Wave: A symmetrical waveform that contains 360 electrical degrees.

AC Voltage: Electrical pressure that reverses direction at regular intervals called cycles.

Absorption: Loss of power in an optical fiber, resulting from conversion of optical power into heat and caused principally by impurities, such as transition metals and hydroxyl ions.

Access Line: A line or circuit that connects a customer site to a network switching center or local exchange. Also known as the local loop.

Aerial Cable: Telecommunications cable installed on aerial supporting structures such as poles, sides of buildings, and other structures.

Alternator: Converts mechanical energy into electrical energy.

Ambient Temperature: The temperature of the surroundings.

American National Standards Institute (ANSI): A private organization that coordinates some U.S. standards setting.

American Standard Code for Information Interchange (ASCII): A standard character set that (typically) assigns a 7-bit sequence to each letter, number, and selected control character.

American Wire Gauge (AWG): Standard used to describe the size of a wire. The larger the AWG number, the smaller (thinner) the described wire.

Ammeter: An instrument (meter) for measuring electrical current.

Ampacity: The amount of current (measured in amperes) that a conductor can carry without overheating.

Ampere (A): Unit of current measurement. The amount of current that will flow through a 1 ohm resistor when 1 volt is applied.

Ampere-hour: The quantity of electricity equal to the flow of a current of 1 ampere for 1 hour.

Amplitude: The size, in voltage, of signals in a data transmission.

Angular Misalignment: A deviation from optimum alignment from fiber to fiber, or fiber to waveguide, resulting in a loss of optical power.

Annunciator: A sound-generating device that intercepts and speaks the condition of circuits or circuits' operations.

Anode: The positive electrode in an electrochemical cell (battery) toward which current flows.

Apparent Power (P$_A$): Product of the voltage and current in a circuit calculated without considering the phase shift that might be present between the voltage and the current. Expressed in terms of volt: ampere (VA).

Approved Ground: A grounding bus or strap in a building that is suitable for connecting to data communication equipment.

Arcing: A luminous discharge formed by the span of electrical current across a space between electrodes or conductors.

Arc Tube: The light-producing element of an HID lamp.

Armature: The rotating part of a DC motor.

Arrestor (lightning): A device that reduces the voltage of a surge applied to its terminals and restores itself to its original operating condition.

Attenuation: Denotes the loss in strength of power between that transmitted and that received. Usually expressed as a ratio in dB (decibel).

Attenuator: A device that reduces signal power in a fiber-optic link by inducing loss.

Autotransformer: Changes voltage level using the same common coil for both the primary and the secondary.

Avalanche Current: Current passed when a diode malfunctions.

Average Value (Vavg): The mathematical mean of all instantaneous voltage values in a sine wave.

B: Byte.

b: bit.

Back Electromagnetic Force: The voltage created in an inductive circuit by a changing current flowing through the circuit.

Back Reflection, Optical Return Loss: Light reflected from the polished end of a fiber caused by the difference of refractive indices.

Backboard: A wooden (or metal) panel used for mounting equipment usually on a wall.

Backbone: The main connectivity device of a distributed system. All systems that have connectivity to the backbone will connect to each other.

Ballast: An electrical circuit component used with fluorescent lamps to provide the voltage necessary to strike the mercury arc within the lamp, and then to limit the amount of current that flows through the lamp.

Bandwidth: Technically, the difference, in hertz (Hz), between the highest and lowest frequencies of a transmission channel.

Bank: An assemblage of fixed contacts.

Battery: A device that converts chemical energy into electrical current.

Baud: A unit of signaling speed. The speed in baud is the number of discrete conditions or signal elements per second.

Bidirectional Current: Has both positive and negative values.

Bits per Second (bps): Basic unit of measurement for serial data transmission capacity, abbreviated as k bps, or kilobit/s, for thousands of bits per second; m bps, or megabit/s, for millions of bits per second; g bits, or gigabit/s for billions of bits per second.

Blocking Diode: A diode used to prevent current flow in a photovoltaic array during times when the array is not producing electricity.

Bonding: A very low impedance path accomplished by permanently joining non-current-carrying metal parts. It is done to provide electrical continuity and to offer the capacity to safely conduct any current.

Bonding Jumper: A conductor used to ensure the required electrical connection between metal parts of an electrical system.

Bonding Conductor: The conductor that connects the non-current-carrying parts of electrical equipment, cable raceways, or other enclosures to the approved system ground conductor.

Bond Wire: Bare grounding wire that runs inside of an armored cable.

Branch Circuit: Conductors between the last overcurrent device and the outlets.

Branch Circuit, Multiwire: A branch circuit having two or more ungrounded circuit conductors, each having a voltage difference between them, and a grounded circuit conductor (neutral) having an equal voltage difference between it and each ungrounded conductor.

Break: The number of separate places on a contact that open or close a circuit.

Breakout Box: A device that allows access to individual points on a physical interface connector for testing and monitoring.

Breakout Cable: A multifiber cable where each fiber is protected by an additional jacket and strength element beyond the overall cable.

British Thermal Unit (Btu): The amount of heat necessary to raise the temperature of 1 pound of water by 1°F.

Brushes: Sliding contacts that provide the connection between the external power circuit and the commutator of a DC motor.

Bus: A group of conductors that serve as a common connection for circuits.

Bus Bar: The heavy copper or aluminum bar used to carry currents in switchboards.

Busway: A metal-enclosed distribution of bus bars.

Cable: One or more insulated or non-insulated wires used to conduct electrical current.

Calorie (Cal): The amount of heat required to raise the temperature of 1 gallon of water by 1°C.

Capacitance (C): The ability of a circuit or component to store an electrical charge and is measured in farads (F).

Capacitive Circuit: A circuit in which current leads voltage (voltage lags current).

Capacitive Reactance (Xc): The opposition to current flow by a capacitor measured in ohms (Ω).

Capacitor: A device that stores electrical energy by means of an electrostatic field.

Cathode: The negative electrode in an electrochemical cell (battery).

Choke Coil: An inductor used to limit the flow of AC.

Circuit: A complete path through which electricity may flow.

Circuit Breaker: A device used to open and close a circuit by automatic means when a predetermined level of current flows through it.

Circular Mil (cm): A measurement of the cross-sectional area of a conductor; kcmil equals 1,000 circular mils.

Closed Circuit: A continuous path providing for electrical flow.

Coil: A winding of insulated conductors arranged to produce magnetic flux.

Common Noise: Noise produced between the ground and the hot lines or the ground and the neutral lines.

Commutator: A series of copper segments connected to the armature of a DC motor.

Conductance: The measure of the ability of a component to conduct electricity expressed in mhos.

Conductor: A substance that offers little resistance to the flow of electrical currents. Insulated copper wire is the most common.

Conduit Body: The part of a conduit system, at the junction of two or more sections of the system, that allows access through a removable cover. Most commonly known as condulets, LBs, LLs, and LRs.

Contacts: The conducting part of a switch that operates with another conducting part to make or break a circuit.

Contactor: A control device that uses a small current to energize or de-energize the load connected to it.

Continuous Load: A load whose maximum current continues for three hours or more.

Core: The central, light-carrying part of an optical fiber.

Coulomb: An electrical current of 1 ampere per second.

Crosstalk: The unwanted energy transferred from one circuit or wire to another which interferes with the desired signal. Usually caused by excessive inductance in a circuit.

Current: The flow of electricity in a circuit, measured in amperes.

Cutout Box: A surface-mounted electrical enclosure with a hinged door.

Cutback Method: A technique for measuring the loss of bare fiber by measuring the optical power transmitted through a long length then cutting back to the source and measuring the initial coupled power.

Cycle: Measured in hertz (Hz), it is the flow of AC in one direction and then in the opposite direction in one time interval completing one wavelength. Hertz is expressed in cycles per second.

Daisy Chaining: The connection of multiple devices in a serial fashion.

Data Rate: The number of bits of information in a transmission system, expressed in bits per second (bps), and which may or may not be equal to the signal or baud rate.

dB: Decibel referenced to a microwatt.

dBm: Decibel referenced to a milliwatt.

Decibel: 1) A standard logarithmic unit for the ratio of two powers, voltages, or currents. In fiber optics, the ratio is power. 2) Unit for measuring relative strength of a signal parameter such as power or voltage.

Deep Cycle: Battery type that can be discharged to a large fraction of capacity. See Depth of Discharge.

Degauss: To remove residual permanent magnetism.

Delta Connection: A connection that has each coil end connected end-to-end to form a closed loop.

Depth of Discharge (DOD): The percent of the rated battery capacity that has been withdrawn.

Device (or wiring device): The part of an electrical system that is designed to carry, but not use, electrical energy.

Diac: A thyristor that triggers in either direction when its breaker voltage is exceeded.

Diode: Electronic component that allows current flow in one direction only.

Direct Current (DC): Electrical current that flows in one direction only.

DC Compound Motor: The field is connected in both series and shunt with the armature.

DC Permanent Magnet Motor: Uses magnets, not a coil, for the field winding.

DC Series Motor: The field is connected in series with the armature.

DC Shunt Motor: The field is connected in parallel with the armature.

DC Voltage: Voltage that flows in one direction only.

Double Break Contacts: Contacts that break the current in two separate places.

Dispersion: Causes a spreading of light as it propagates through an optical fiber. Its three types are modal, material, and waveguide.

Dual Voltage Motor: A motor that operates at more than one voltage level.

Eddy Current: Unwanted current induced in the metal field structure of a motor.

Effective Current: The value of AC that causes the same heating effect as a given value of DC. For sine wave AC, the effective current is 0.7071.

Effective Value: Also called the root-mean-square (RMS), it produces the same I^2R power as an equal DC value.

Efficiency (EFF): The ratio of output power to input power, usually expressed as a percentage.

Electricity: The movement of electrons through a conductor.

Electromagnet: Coil of wire that exhibits magnetic properties when current passes through it.

Electron: The subatomic unit of negative electricity expressed as a charge of 1.6×10^{-19} coulomb.

Electronics: The science of treating charge flow in vacuums, gases, and crystal lattices.

Energy: The capacity to do work.

Equalization: The process of restoring all cells in a battery to an equal state of charge.

Ethernet: A 10-Mbps, coaxial standard for LANs. All nodes connect to the cable where they contend for access via CSMA/CD.

Excess Loss: In a fiber-optic coupler, the optical loss from that portion of light that does not emerge from the nominally operational ports.

Excitation: The power required to energize the magnetic field of transformers, generators, and motors.

Extrinsic Loss: In a fiber interconnection, that portion of loss that is not intrinsic to the fiber but is related to imperfect joining, which may be caused by the connector or splice.

Farad (F): The unit of measurement of capacitance.

Fault Current: Any current that travels a path other than the normal operating path of an electrical system.

Feeder: Circuit conductors between the service and the final branch circuit overcurrent device.

Ferrule: A precision tube that holds a fiber for alignment for interconnection or termination.

Fiber Optics: A technology that uses light as a digital information carrier.

Field: The stationary windings (magnets) of a DC motor.

Filament: A conductor that has a high enough resistance to cause heat.

Filter: Circuit elements combined specifically to pass certain frequencies and resist all others.

Flashover: A disruptive electrical discharge around or over (but not through) an insulator.

Fluorescence: The emission of light by a substance when exposed to radiation or the impact of particles.

Flux: An electrical energy distributed in space and represented diagrammatically by means of flux lines denoting magnetic or electrical forces.

Footcandle (fc): The amount of light produced by a lamp measured in lumens divided by the area that is illuminated.

Four-Wire Circuits: Telephone circuits that use two separate one-way transmission paths of two wires each, as opposed to regular local lines that usually only have two wires to carry conversations in both directions.

Frequency: The number of times per second a signal regenerates itself at a peak amplitude. It can be expressed in hertz (Hz), kilohertz (kHz), megahertz (MHz), etc.

Full-Load Current (FLC): The current required by a motor to produce the full-load torque at the motor's rated speed.

Full-Load Torque (FLT): The torque required to produce the rated power of the motor at full speed.

Fuse: A protective device, also called an OCPD, with a fusible element that opens the circuit by melting when subjected to excessive current.

Gain: Ratio of the amplitude of the output signal to the input signal.

Gap Loss: Loss resulting from the end separation of two axially aligned fibers.

Gb: Gigabit. 1 billion bits of information.

Gbyte: Gigabyte. 1 billion bytes of data.

Ghost Voltage: A voltage that appears on a motor that is not connected.

Grid: Term used to describe an electrical utility distribution network.

Ground: An electrical connect (on purpose or accidental) between an item of equipment and the earth.

Grounding: The connection of all exposed non-current carrying metal parts to the earth.

Ground Fault: A condition in which current from a hot power line is flowing to the ground.

Guy: A wire having one end secured and the other fastened to a pole or structure under tension.

Ground Fault Circuit Interrupter (GFCI): An electrical device that protects personnel by detecting hazardous ground faults and quickly disconnecting power from the circuit.

Harmonic: 1) A sinusoid with a frequency that is an integral multiple of a certain frequency. 2) The full multiple of a base frequency.

Heater: A device that is placed in a motor starter to measure the amount of current in the power line.

Heat Sink: A piece of metal used to dissipate the heat of solid-state components mounted on it.

Heating Element: A conductor (wire) that offers enough resistance to produce heat when connected to power.

Henry (H): The unit of measure of inductance in which a current changing its rate of flow 1 ampere per second induces an electromotive force of 1 volt.

Hertz (Hz): The unit of measure of the frequency of the AC sine wave to complete a cycle; 1 Hertz is equal to one cycle of the AC sine-wave per second.

HID Lamp: High Intensity Discharge Lamp.

Holding Current: The minimum current necessary for an SRC to continue conducting.

Horsepower (HP): A unit of power equal to 746 watts that describes the output of electric motors.

Hub: A device that connects to several other devices usually in a star topology. Also called concentrator, multiport repeater or multi-station access unit (MAU).

Hybrid: An electronic circuit that uses different cable types to complete the circuit between systems.

Impedance (Z): The total opposition offered to the flow of AC from any combination of resistance, inductive reactance, and capacitive reactance and is measured in ohms (Ω).

Inductance (L): The property of a circuit that determines how much voltage will be induced into it by a change in current of another circuit and is measured in henrys (H).

Inductive Circuit: A circuit in which current lags voltage.

Inductive Reactance (X_L): The opposition to the flow of AC in a circuit due to inductance and is measured in ohms (Ω).

In-Phase: The state when voltage and current reach their maximum amplitude and zero level at the same time in a cycle.

Insertion Loss: The loss of power that results from inserting a component, such as a connector or splice, into a previously continuous path.

Insulator: Material that current cannot flow through easily.

Integrated Circuit or IC: A circuit in which devices such as transistors, capacitors, and resistors are made from a single piece of material and connected to form a circuit.

Interface: The point where two systems, with different characteristics, connect.

Isolated Grounded Receptacle: Minimizes electrical noise by providing a separate grounding path.

Isolation Transformer: A one-to-one transformer that is used to isolate the equipment at the secondary from earth ground.

Jack: A receptacle (female) used with a plug (male) to make a connection to in-wall communications cabling or to a patch panel.

Jacket: The protective and insulating outer housing on a cable. Also called a sheath.

Jogging: The frequent starting and stopping of a motor.

Joule: A unit of electrical energy also called a watt-second; the transfer of 1 watt for 1 second.

Jumper: Patch cable or wire used to establish a circuit, often temporarily, for testing or diagnostics.

Junction Box: A box, usually metal, that encloses cable connections.

KB: Kilobyte (1,000 bytes).

Kb: Kilobit (1,000 bits).

Kbps: Kilobits per second (1,000 bits per second).

kcmil: 1,000 circular mils.

kVA: Kilovolt-amperes (1,000 volt amps).

kVAR: Kilovar (1,000 reactive volt amps).

kV: Kilovolt (1,000 volts).

kW: Kilowatt (1,000 watts).

kWH: Kilowatt hour: The basic unit of electrical energy for utilities equal to a 1,000 watts of power supplied for 1 hour.

Lamp: A light source. Reference is to a light bulb, rather than a lamp.

Leakage Current: Current that flows through insulation.

Leg (circuit): One of the conductors in a supply circuit in which the maximum voltage is maintained.

Load: The amount of electric power used by any electrical unit or appliance at any given moment.

Location, Damp: Partially protected locations, such as under canopies, roofed open porches, etc. Also, interior locations that are subject only to moderate degrees of moisture, such as basements, barns, etc.

Location, Wet: Locations underground, in concrete slabs, where saturation occurs, or outdoors.

Locked Rotor: Condition when a motor is loaded so heavily that the shaft cannot turn.

Locked Rotor Current (LRC): The steady-state current taken from the power line with the rotor locked and the voltage applied.

Locked Rotor Torque (LRT): The torque a motor produces when the rotor is stationary and full power is applied.

Loss Budget: The amount of power lost in a fiber optic link. Used in terms of the maximum amount of loss that can be tolerated by a given link.

Loss, optical: The amount of optical power lost as light is transmitted through fiber, splices, couplers, and the like.

Lumen (lm): The unit used to measure the total amount of light produced by a light source.

Mbps: Million bits per second.

Mbyte: Megabyte. Million bytes of information.

Magnetic Field: The invisible field produced by a current-carrying conductor or coil, permanent magnet, or the earth itself, that develops a north and south polarity.

Magnetic Flux: The invisible lines of force that make up the magnetic field.

Magnetic Induction: The setting up of magnetic flux lines in a material by an electric current. The number of lines is measured in maxwells.

Maxwell: The unit of measurement of the total number of magnetic flux lines in a magnetic field.

Mechanical Splice: A semipermanent connection between two fibers made with an alignment device and index-matching fluid or adhesive.

Micron (m): A unit of measure, 10^{-6} m, used to measure wavelength of light.

Mode: A single electromagnetic field pattern that travels in fiber.

Motor: A machine that develops torque (rotating mechanical force) on a shaft to produce work.

Motor Efficiency: The effectiveness of a motor to convert electrical energy into mechanical energy.

Motor Starter: An electrically operated switch (contactor) that includes overload protection.

Motor Torque: The force that produces rotation in the shaft.

Multiplex: To combine multiple input signals into one for transmission over a single high-speed channel. Two methods are used: (1) frequency division, and (2) time division.

Mutual Inductance: The effect of one coil inducing a voltage into another coil.

Nanometer (nm): A unit of measure, 10^{-9} m, used to measure the wavelength of light.

NEC®: National Electrical Code®, which contains safety rules for all types of electrical installations.

No-Load Current: The current demand of a transformer primary when no current demand is made on the secondary.

Normally Open Contacts: Contacts that are open before being energized.

Normally Closed Contacts: Contacts that are closed before being energized.

Ohm: The unit of measurement of electrical resistance; 1 ohm of resistance will allow 1 ampere of current to flow through a pressure of 1 volt.

Ohm's Law: A law that describes the mathematical relationship between voltage, current, and resistance.

Open Circuit: A condition that provides no path for electric current to flow in a circuit.

Open Circuit Voltage: The maximum voltage produced by a photovoltaic cell, module, or array without a load applied.

Optical Power: The amount of radiant energy per unit of time, expressed in linear units of watts or on a logarithmic scale, in dBm (where 0 dB = 1 mW) or dB (where 0 dB = 1 W).

Oscillation: Fluctuations in a circuit.

Out of Phase: Having AC sine waveforms that are of the same frequency but are not passing through corresponding values at the same instant.

Outlet: The place in the wiring system where the current is taken to supply equipment.

Overcurrent: Too much current.

Overload Protection: A device that prevents overloading a circuit or motor such as a fuse or circuit breaker.

Parallel Circuit: Contains two or more loads and has more than one path through which current flows.

Peak Value: The maximum value of either the positive or negative alteration of a sine wave.

Period: The time required to produce one cycle of a waveform.

Phase: The fractional part of a period through which the time variable of a periodic quantity has moved.

Phase Converter: A device that derives three-phase power from single-phase power.

Phase Shift: The state when voltage and current in a circuit do not reach their maximum amplitude and zero level at the same time.

Photovoltaic: Changing light into electricity.

Pigtail: A short length of fiber permanently attached to a component.

Polarity: The particular state of an object, either positive or negative, that refers to the two electrical poles, north and south.

Power (watts): A basic unit of electrical energy, measured in watts.

Power Budget: The difference (in dB) between the transmitted optical power (in dBm) and the receiver sensitivity (in dBm).

Power Factor: The ratio of true power (kW) to apparent power (kVA) for any given load and period of time.

Primary Winding: The coil of a transformer that is energized from a source of alternating voltage and current. The input side.

Pulse Spreading: The dispersion of an optical signal with time as it propagates through an optical fiber.

Reactance (X): The measure of the induced voltage due to the inductance of a circuit and is measured in ohms (Ω).

Relay: A device that controls one electrical circuit by opening and closing the contacts in another circuit.

Resistance (R): The opposition to the flow of current in an electrical circuit and is measured in ohms (Ω).

Resistive Circuit: A circuit containing resistive loads such as heating elements or incandescent lamps.

Resonance (f): A condition in a circuit where the frequency of an externally applied force equals the natural tendency of the circuit, when the inductive reactance (X_L) equals capacitive reactance (X_C). Resonance is measured in hertz (Hz).

RG-58: The coaxial cable used by Thin Ethernet (10Base2). It has a 50 ohm impedance and so must use 50 ohm terminators.

RG-59: The coaxial cable, with 75 ohm impedance, used in cable TV and other video environments.

RG-62: The coaxial cable, with 93 ohm impedance, used by ARCNet and IBM 3270 terminal environments.

RJ11: A standard six-conductor modular jack or plug that uses two to six conductors.

RJ22: The standard four-conductor modular jack that connects a telephone handset to its base unit.

RJ45: A standard eight-conductor modular jack or plug that uses two to eight conductors.

Root-Mean-Square: A mathematical expression equal to 0.707 times the peak value of an AC waveform.

Rotary Convertor: A type of phase convertor.

Rotor: The rotating part of an AC motor.

Secondary Winding: The winding of a transformer that receives electrical energy by electromagnetic induction from the primary winding and then delivers it to a load. The output side of a transformer.

Series Circuit: A circuit that has only one current path to and from the power source.

Short Circuit: An undesired path for electrical current.

Shunt: Denotes a parallel connection.

Sine Wave: A waveform corresponding to a single-frequency, periodic oscillation, which can be shown as a function of amplitude against angle and in which the value of the curve at any point is a function of the sine of that angle.

Single-Phase Power: One of the three alternating currents in a circuit.

Solenoid: An electromagnet with a movable iron core.

Split-Phase Motor: A single-phase AC motor that has a running and a starting winding.

Split-Wired Receptacle: A receptacle that has the metal tap removed between the hot terminals.

Stator: The stationary part of an AC motor.

Switch (electrical): A device to start or stop the flow of electricity.

Surge Capacity: The requirement of an invertor to tolerate a momentary current surge imposed by starting AC motors or transformers.

T1: A digital carrier facility for transmitting a single DS1 digital stream over two pairs of regular copper telephone wires at 1.544 Mbps.

T2: A digital carrier facility used to transmit a DS2 digital stream at 6.312 Mbps.

T3: A digital carrier facility used to transmit a DS3 digital stream at 44.746 Mbps.

T4: A digital carrier facility used to transmit a DS4 digital stream at 273 mbps.

Taps: Connecting points on a transformer coil.

Thermal Protection: An electrical device that has inherent protection from overheating. Typically in the form of a bimetal strip that bends when heated to a certain point.

Three-Phase Power: A combination of three alternating currents (usually denoted as a, b, and c) in a circuit with their voltages displaced $120°$ or one-third of a cycle.

Transformer: A device that uses magnetic force to transfer electrical energy from one coil of wire to another. In the process, transformers can also change the voltage at which this electrical energy is transmitted.

True Power (P_T): The actual power used in an electrical circuit measured in watts (w) or kilowatts (kW).

Turns Ratio: The ratio between the voltage and the number of turns on the primary and secondary windings of a transformer.

VA: Volt-amperes.

Volt (E) or (V): The unit of measurement of electrical pressure (force); 1 volt will force 1 ampere of current to flow through a resistance of 1 ohm.

Voltage Drop: Voltage reduction due to wire resistance.

Watt (W): The unit of measurement of electrical power produced by a current of 1 amp across a potential difference of 1 volt.

Waveform: Characteristic shape of an electrical current or signal.

Wavelength: The distance between the same two points on adjacent waves; the time required for a wave to complete a single cycle.

Wye Connection: A connection that has one end of each coil connected together and the other end of each coil left open for external connections.

About the Author

Paul Rosenberg has an extensive background in the electrical, data, construction, and fiber optic trades. He is a leading voice in the electrical industry with years of experience from apprentice to project manager. Paul has written for all of the leading electrical and low-voltage industry magazines and has authored more than 60 books.

In addition, he wrote the first standard for the installation of optical cables in buildings (ANSI-NEIS-301) and was awarded a patent for a power transmission module. Paul has taught 19 continuing education courses for Iowa State University and works as a consultant and expert witness in legal cases. He speaks occasionally at industry events.